International Acclaim for Tim Flannery's *The Eternal Frontier:*

"Bold . . . Flannery's chronicle of evolutionary triumphs and setbacks, foreign invasions, and the rise of some continental specialties—such as horses, camels, dogs, and even cheetahs—would alone make the book worth reading. . . . Watch out, fellow North Americans. This gutsy Aussie may have read our landscape and ecological history with greater clarity than any native son."   —David A. Burney, *Natural History*

"A sweeping natural history of North America from its birth as a self-contained continent in the Cretaceous Era to its current precarious stat as an ecological superpower . . . Natural history par exce
          —*Kirkus R*

"Flannery has a talent for imagining prehistoric land   bringing them vividly to life. . . . A fascinating, current, and insightful look at our familiar history from a larger perspective."
          —David Bezanson, *The Austin American-Statesman*

"Riveting . . . A masterpiece of modern science . . . *The Eternal Frontier* is nature poetry and cautious science, both at their very best."
          —*The Sunday Telegraph* (London)

"Splendid . . . Flannery has been compared to those other popularizers of historical biology, Stephen Jay Gould and Jared Diamond. I think he is superior. He is a better writer than either. . . . Tim Flannery is the real thing: a man with a gift for lucid exposition, who can really make his subject come alive."   —Frank McLynn, *The Literary Review*

The Eternal Frontier

OTHER BOOKS BY THE AUTHOR

Mammals of New Guinea

Tree Kangaroos: A Curious Natural History with R. Martin, P. Schouten and A. Szalay

The Future Eaters: An Ecological History of the Australasian Lands and Peoples

Possums of the World: A Monograph of the Phalangeroidea with P. Schouten

Mammals of the South West Pacific and Moluccan Islands

Watkin Tench, 1788 (ed.)

The Life and Adventures of John Nicol, Mariner (ed.)

Throwim Way Leg

The Explorers (ed.)

The Birth of Sydney

Terra Australis: Matthew Flinders' Great Adventures in the Circumnavigation of Australia (ed.)

A Gap in Nature with P. Shouten

The
Eternal
Frontier

An ecological history of North America
and its peoples

Tim Flannery

GROVE PRESS
New York

First published in 2001 by The Text Publishing Company, Melbourne, Victoria, Australia

ILLUSTRATION SOURCES
Grateful acknowledgment is made to the following for permission to reproduce the illustrative material in the picture sections:
Plate I, painted by Peter Schouten; plate II, Wilbur Garrett/NGS Image Collection; plates III, IV, V and VI, painted by Peter Schouten; plate VII, David Arnold/NGS Image Collection; plate VIII, Michael Nichols/NGS Image Collection; plate IX, University of Alaska; plate X, George Grall/NGS Image Collection; plate XI, Photo Archives, Denver Museum of Nature and Science; plate XII, Library of Congress; plate XIII, Dept. of Library Services, American Museum of Natural History, neg. no. 2A22680; plate XIV, Arizona Historical Society/Tucson, AHS 20602; plate XV, National Cowboy Hall of Fame and Western Heritage Center, Oklahoma City; plate SVI, Burton Historical Collection, Detroit Public Library; plate XVII, from *Audubon's Birds of America,* La Trobve Collection, State Library of Victoria; plate XVIII, painted by Peter Schouten.

Published simultaneously in Canada
Printed in the United States of America

Library of Congress Cataloging-in-Publication Data

Flannery, Tim F. (Tim Fridtjof), 1956–
    The eternal frontier : an ecological history of North America and its peoples / Tim Flannery.
        p. cm.
    First published by Text Publishing, Melbourne, Australia, in 2001.
    ISBN 0-8021-3888-8 (pbk.)
    1. Natural history—North America. 2. Ecology—North America. I. Title.

    QH102 .F63 2001
    508.7—dc21   2001018841

Design by Chong Weng-ho
Maps by Norman Robinson

Grove Press
841 Broadway
New York, NY 10003

02 03 04 05   10 9 8 7 6 5 4 3 2

This project has been assisted by the Government of South Australia through Arts South Australia and by the Commonwealth Government through the Australia Council, its arts funding and advisory body.

To the North Americans:
in admiration of the efforts so many are now making
to win back the natural grandeur, the biodiversity
and ecological balance of their exceptional land.

CONVERSION TABLE

LENGTH

1 centimetre = 0.39 inches

1 metre = 3.28 feet

1 kilometre = 0.62 miles

MASS

1 kilogram = 2.2 pounds

1 tonne = 0.98 ton

AREA

1 hectare = 2.47 acres

1 square kilometre = 0.39 square miles

VOLUME

1 cubic kilometre = 0.24 cubic miles

TEMPERATURE

$°F = (°C \times 1.8) + 32$

Contents

Act 2

IN WHICH AMERICA BECOMES A TROPICAL PARADISE

---

4 FIRST CONTACTS

Mr Stein's tiny fossil. The great uintathere war. Buffalo Bill meets the professor. Fossil hunting on the frontier. Iguanas become the first wetbacks. Ancient invaders from Asia, including the hedgehog that became a hippo.

5 THE BRIDGE OVER GREENLAND

A global heatwave and its consequences. Aussie turtles and goannas take over. The opening, and closing, of the bridges across Greenland. North America invades Europe. Asia invades North America—or does it? How Europe's camels came to look like rabbits. The curse of knowing too much. Size really does count, especially east of Eden. The beginnings of the modern world order.

Act 3

IN WHICH AMERICA BECOMES A LAND OF IMMIGRANTS

---

6 A FATAL CONFIGURATION

North America the great amplifier. The continent as a trumpet with an ear to the weather. A discourse on whether North America is one or two places in space and time, and on the importance of such a perception. An inquiry into nuts, squirrels, leaves like stars and other strange phenomena.

falling seas and the land itself heaves. Canadian forests find refuge in
Florida, while prairie herbs creep under coniferous trees. A mountain
highway saves the navy.

12 VISIT TO A NEW WORLD
The final flowering of a unique fauna. An ice-age bestiary. Jumbo
elephants, giant killer bears, a plethora of pussies, new world cheetahs
and so much more. The feathered legions and an avian impossibility. Last
relics haunt a frozen north and bright Californian skies.

Act 4

IN WHICH AMERICA IS DISCOVERED

---

13 A NEW WORLD
The dating game and the pre-Clovis phantom. Which barrier? Alaska the
revolving door. Daniel Boone of the Pleistocene and the first frontier
experience. The forgetting of art? Deadly sculptures in flint.

14 THE BLACK HOLE THEORY OF EXTINCTION
Mr Wallace's marvellous fact, and the time of pygmies. Who died where,
when and why? Climate, the spear, or 'racial senility'? Global killing and
Australia as the great separate experiment. The Cuban anomaly.

15 MASSACRING THE MAMMOTH, DISMEMBERING THE MASTODON
It's hard to kill an elephant. The sharp eyes of Ed Lehner.
Dr Laub and the dear deceased pachyderm. The pond near Hell, or is
lamb like mammoth? The living saint of mammoth taphonomy.
The last Clovis point?

Act 5

IN WHICH AMERICA CONQUERS THE WORLD

peasants, soil and furs—the Spanish, French and English
experience in America.

21 ENGLISH COLONIES ALL

Of pilgrims and penury. The plague that saved the plantation. A ruinous
fortune. Sex and sentencing in old Boston. A failed frontier. The tyranny
of tobacco and how a pioneer weed created a pioneer culture. The first
agricultural revolution since Rome. Topics for the evolutionist.
Union and the first American civil war.

22 CONCEIVED IN LIBERTY

Make way for the young American buffalo. The amazing Frederick
Jackson Turner and the American frontier. Making North Americans. The
return of the native. Is the peccary a hog? Prince Colloredo Mannsfeld
and his love of small furry things. The ass and the bighorn. Of killer
funguses and woolly adelgids.

23 THE FATAL IMPACT

From garefowl to barn-door skate—a sorry tale of overeating. Merciful
pilgrims and murderous microbes. A New World pox? First germ warfare.
Apache get your gun. Three hundred and seventy treaties later.

24 AMERICA UNDER THE GUN

A bird for a young nation. Canned curlew anyone? The Boswell
of the bison. The virtues of buffalo fertiliser, and how the prairie
paid the ultimate buffalo bill. Hens and humans replace
flocks and herds.

25 THE MAKING OF A GIANT

The first billionaire. The unholy trinity of transport, industry and unity.
Making guns gives more power than using them. A willingness
to innovate. Mr Ford and his beloved soybean. Bonanza

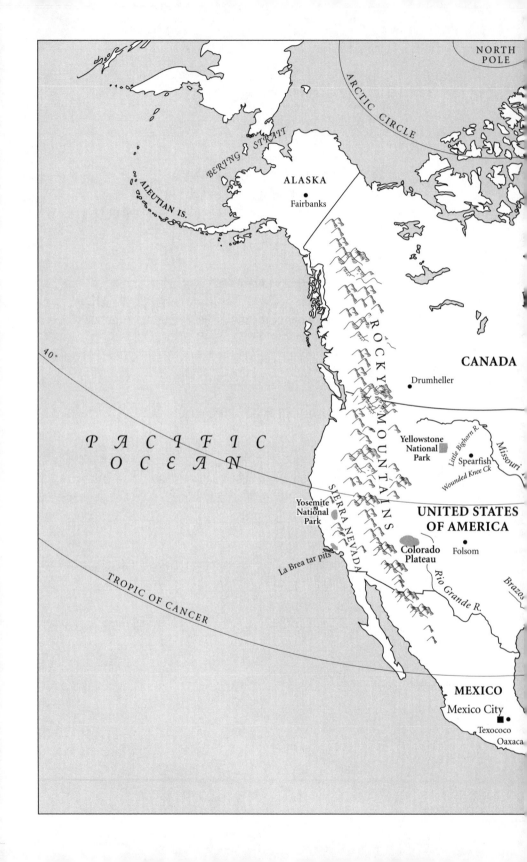

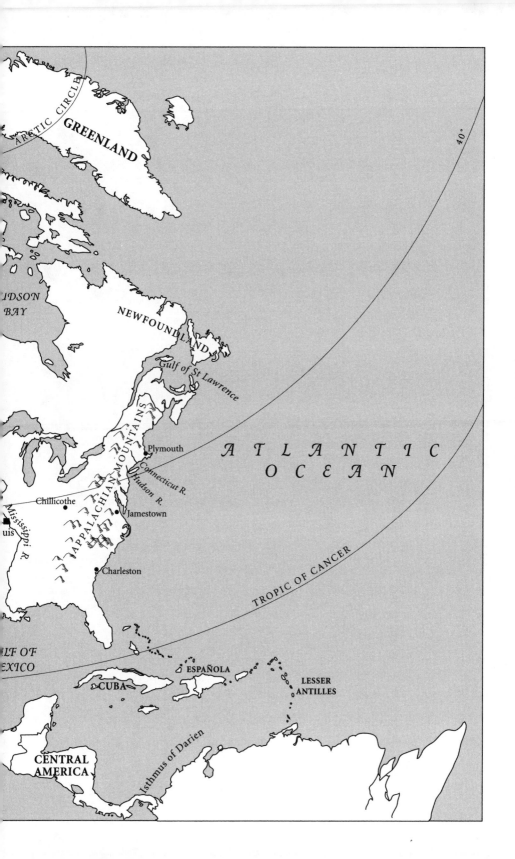

INTRODUCTION

There are forces in the lives of people, and animals and plants too, that have made them what they are. Some are big forces—factors of climate and topography, isolation and landmass size—that exert their power for eons and in the process shape species and whole environments. Others are more transient and affect populations, communities and individuals as circumstances change rapidly around them. Ecological historians range far and wide through time and space, rounding up these disparate forces to show the evolution of a particular species or an entire region.

North America seems too big and too diverse to be easily comprehended. To visitors and even to many of its inhabitants it remains something to be marvelled at—an enigma. This history seeks to identify the forces that have shaped, and the common threads that run through, the experiences of life in the great continental drama that has been played out in North America over the last 65 million years. I hope to provide a lens through which North America's biota—animal, plant and human—as well as the grand landscape itself, can be viewed afresh.

As I began to write I found myself asking, what *is* North America? The question has no easy answers, even in geographic terms, for the continent has changed markedly over time. I found that Alaska (in a pattern that reflects its recent political past) has often been connected with Asia and divided from the rest of North America. Should it, along with tentatively connected Greenland, be considered part of North America for the purposes of an ecological history? Most vexing was the question of where North America ends and Central or South America begins. Ichthyologists, botanists, linguists and political scientists would place the border at or near the Rio Grande, while geographers, geologists and some economists would have it closer to Panama. The great Charles Darwin himself divided the Americas 'not by the Isthmus of Panama, but by the southern part of Mexico, in Latitude 20°'. Eventually I accepted the definition of North America used by geographers, bounding the continent at the Bering Strait and the Isthmus of Darien. Yet the further one travels back in time the less relevant are definitions based on contemporary geography, for North America has had a particularly turbulent and dynamic past.

North America is the middle child in the family of habitable continents. The southern realms of Australia and South America are smaller, while Eurasia and Africa are larger. Despite its 'middleness', it is an exceptional place, for it is home to breathtaking natural diversity, including such marvels as the globe's tallest and oldest trees, the Californian condor and lesser known wonders such as the Mexican burrowing toad. It is also possessed of what is arguably the best fossil record of any continent, which reveals in detail how this land has changed through time and space, gradually becoming what it is today.

A Renaissance European way of thinking has saddled us with a globe divided conceptually in two—an Old and a New World. Despite the fact that neither is a biologically coherent entity, scientists have formalised the concept, designating the landmasses of the eastern hemisphere as the Old, and of the western hemisphere as the New World. But there is another sense in which North America is a new world, for time and again during its history it has been discovered by living things and has acted as their expansive frontier. From rattlesnakes to hummingbirds and bison, the frontier experience has changed them all, each time creating a new and distinctive manifestation of life.

North America was the last but one of the habitable continents to feel the footprint of humanity. Colonised by Clovis hunters around 13,200 years ago, by 1492 it had become home to an extraordinary diversity of cultures, from highly organised nation states to hunter-gatherers, including some unique forms of human social organisation. Since the eighteenth century it has been the continent of the historic frontier, of Henry Thoreau and of Henry Ford, and today it is home to the sole remaining political superpower.

It is only sensible to write ecological histories about ecologically coherent entities, and at the beginning of this project I was not convinced of the ecological coherence of North America. Its enormous complexity left me wondering whether a common set of environmental determinants could produce such dissimilar phenomena. I could not see how the sublime New England fall and Sonoran cacti could be shaped by a common evolutionary force. I worried over the fact that the US east and west are so different—redwoods and medicinally sanctioned marijuana in California, codfish and Puritanism in New England. Why, I puzzled, are native North American cultures so diverse and how, if at all, are contemporary American societies being shaped by their environment?

As a result of writing this book I now believe that life in North America has been shaped by a unique combination of forces generated by the continent's singular geography, geological history and climate. It is these powerful forces that have produced what is distinctive about

North America, and they have profoundly affected the experiences of the continent's immigrants and emigrants.

Since the arrival of the Europeans North America has been divided into three principal political entities of Hispanic, British and French origin. Each has acted as an experiment in adaptation, finding its own frontier and shaping its fate accordingly on the new continent. While none represents a truly independent experiment, and within each enormous diversity exists, these three influences offer us the chance to examine anew the interactions between a culture and the land.

Although the citizens of the United States of America are colloquially known as Americans, the United States is not synonymous with North America; both Canadians and Mexicans offer very different examples of what being American can mean. Nevertheless, I believe that the history of the US is in some ways an exemplar—the example of examples—of a story recapitulated countless times in North America over the past 65 million years. Abraham Lincoln described his nation as having been 'conceived in liberty'. It was a sentiment born on the frontier and spoken by the most famous frontiersman of them all. For me these three words capture the very essence of the United States of America, and I hope to show they have an ecological as well as a political dimension.

In other ways the US is an anomaly in North American ecological history. The present world economy is an invention of the Americans. US experimentation and reorganisation of the means of production would in time lead to the development of what one economist has called 'a perpetual cornucopia machine'.[1] For all the wealth it created, it's a machine that has stripped North America of much of its natural grandeur; and there is nothing 'perpetual' about it. Its dreadful ecological impact left me pondering Lincoln's Gettysburg address afresh and wondering whether 'a nation so conceived and dedicated, can long endure'.

The great questions for me are these: what are the quintessential determinants of life in North America? Have they remained stable through time, and how strong are they in shaping flora, fauna and human societies? Has North America always been a global cornucopia, or is it only under special circumstances that some of its productions—both human and non-human—can successfully exert a global influence? And can its leading nation, the United States, 'long endure' in its current form?

The roots of causality in North America are profound, and to address these questions we must go far back in time, to when modern North America came into being. That continent-defining moment occurred one balmy day 65 million years ago, give or take 120,000 years, when a great fiery ball appeared in the sky and came crashing to Earth.[2]

What changed North America—and indeed the world at that instant—was an asteroid. The rock had been travelling through space at 90,000 kilometres an hour since time immemorial. Then the statistically improbable happened. It began approaching our planet on a collision course, but where would it hit? Ground Zero, as it happened, was to be North America. This fact, this utterly random event, would change world history.

The scene opens on a continent not conceived in liberty but riven in two, and soon to be visited by fire, flood and famine.

Act I

•

In Which America Is Created and Undone

GROUND ZERO

The death-dealing visitor appeared in the skies 65 million years ago above a planet that had long rested in contented stability. The era it terminated—the age of dinosaurs—had dawned following a crisis in life history some 180 million years earlier. The steady unfolding of evolutionary change in this tranquil world had seen the first birds, first mammals and first flowering plants come into existence. Thus, despite the fact that it was dominated by dinosaurs, the age was the nursery of our times.

At its dawn all the world's continents were joined into one vast landmass known as Pangaea, which was inhabited by many kinds of animals and plants. Pangaea was destined to divide into two super-continents, known as Laurasia and Gondwana, and by the end of the era these landmasses had begun to fragment, giving rise to the contemporary continents.

Continents are not born like people, for there is no day, century or even millennium that we can hail as a continental 'birthday'. This is because continents are born of inexorable geological forces that move at glacial speed. They coalesce, break apart and re-collide, driven by great convection currents circulating in the Earth's mantle. Most of the existing continents were formed of fragmentation: Australia, Antarctica, South America and Africa all came into existence as a result of the break-up of the supercontinent Gondwana. North America, however, was created differently—it resulted from a victory of the forces of union.

North America was born in the twilight moments of the age of dinosaurs, and at birth it was a very different place, for there was no Mississippi River, no Rocky Mountains and no Isthmus of Darien. Instead, a vast shallow seaway, dubbed by Canadian scientists the Bearpaw Sea, occupied its southern and central portions, dividing North America into separate eastern and western landmasses. The Bearpaw Sea had been in existence for at least 35 million years and had acted as a zoogeographic barrier.[1] During their long separate histories the two islands destined to form North America had become as different as Australia and South America are today. Some species did, however, manage to cross the Bearpaw, possibly by using volcanic islands (which would have been important breeding grounds for species such as the giant marine reptiles, seabirds and perhaps pterodactyls) as stepping stones. As the Bearpaw Sea narrowed such crossings became more frequent, heralding the amalgamation of the two island realms into a continental whole.

The landmass that constituted the eastern island of this proto-North America had long been geologically stable and bereft of mountain-building heat and energy. The Appalachian Mountains that were its backbone began to form over 450 million years ago and had already been eroded into a flat-topped cordillera by 230 million years ago. Zoogeographers can identify various animal groups that probably first evolved on this ancient island in the Mesozoic sea, among them the largest family of salamanders in the world, the Plethodontidae or lungless

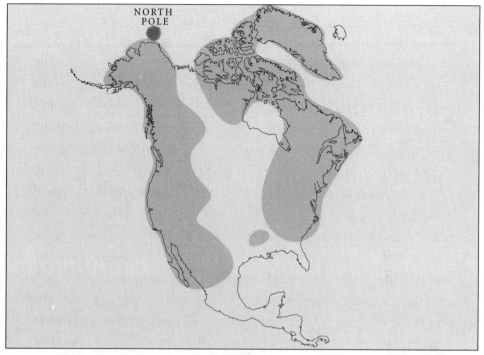

Throughout much of the Cretaceous period North America existed as two isolated islands, separated by the broad but shallow Bearpaw Sea.

salamanders.[2] By the end of the age of dinosaurs the east coast of this ancient land had been eroding and subsiding into the sea for tens of millions of years.

America's western island was, in contrast, a geologically restless land that had been joined to Asia via the Beringian land bridge for hundreds of millions of years. Its fauna and flora were largely shared with Asia. Just before the asteroid struck, the Sierra Nevada had come into existence and was a high, jagged range of recently extinct volcanoes. Rivers draining eastward from their peaks meandered across a broad coastal plain towards the Bearpaw Sea. Looking from those high peaks to the west, one would have seen the Pacific Ocean, but the western coastline was then very different. Where beaches and headlands stand today there was nothing but water, for there was no Baja California (the peninsula was still stitched to Mexico proper) and parts of what are now coastal

Oregon and Washington were islands positioned far out to sea. A portion of what is now coastal California was also islands, situated well south of its present location. Furthermore, most of Mexico was submerged under a shallow sea.

Long before the asteroid struck pressures had been building deep within the Earth's crust that would eventually thrust the Rocky Mountains high into the air, but for the present the movements caused by these pressures served only to narrow the inland seaway that divided America. By the end of the age of dinosaurs its southern section was still a shallow gulf, its mouth located near present-day Houston. In polar regions, however, the seaway changed into freshwater lakes and swamps, providing partial access across the newly unified continent.

Dramatic changes in the vegetation of North America had been occurring as the Bearpaw Sea narrowed. Earlier, great tropical conifers such as relatives of the monkey puzzle tree, which today have a primarily southern hemisphere distribution, had been dominant. At this time the pines of the boreal north, including the family to which the genus *Pinus* belongs, were yet to take on their modern form. Their ancestors were still growing as shrubs in ancient heath-like communities—pygmy plants revealing no signs of their greatness to come. Tens of millions of years would pass before they would come to resemble modern pines. The flowering plants likewise then lived as humble shrubs and twining plants in gaps in the canopy.

As the age of dinosaurs drew to a close many flowering plants began to grow larger and some, such as members of the magnolia family, would become the first flowering trees. Just two million years before the asteroid struck, these trees had formed a great belt of evergreen tropical forest that extended in a swath, south of 60 degrees latitude (the present latitude of Nunivak Island, Alaska), right across the continent's middle. This forest did not entirely replace the great conifers, however, for some were now gigantic trees, towering above the canopy of broadleaved flowering plants. Growth rings are difficult to detect in the fossilised timber of these trees, which indicates an absence of seasons.

To understand the nature of that ancient forest it helps to walk the badlands along the Red Deer River in Alberta, Canada. There, amid the grey, black and tan buttes of soft eroding rock that jut out throughout the valley, the Midland Coal Company found an entire ancient forest turned to coal, which they mined until the 1950s. Even where there are no seams of coal the rocks are packed with ancient plant remains. Leaves of exotic Asian species such as the ginkgo and the dawn redwood cram the sediments, alongside the tropical forms of the araucaria and palm. In one special layer the plant remains were not turned to coal, but were instead replaced with black silica. When freed by acid from the surrounding ironstone these 65 to 70-million-year-old fossils look like the most exquisite jewels—tiny branchlets, seeds and cones preserved in the finest glassy detail. To pick up a dusty rock packed with such fossils from among the sparse badlands shrubs struggling to survive is a strong reminder of just how much the world has changed since the demise of the dinosaurs.

As the continent was thus being born a bright, star-like object appeared in the sky. This was no guiding star but a stony, malevolent extraterrestrial visitor. The death-dealing lump of rock that shone so brightly was at least ten kilometres in diameter. That's a piece of celestial real estate two to three times the length of Australia's Uluru (formerly Ayers Rock), the largest monolith on our planet. We can imagine it silently approaching a steadily revolving Earth; a jagged lump of stone playing a game of Russian roulette with life's mothership. Perhaps it had missed narrowly on a billion earlier occasions, or perhaps it was on a new and deadly trajectory, pulled from its timeless round in the asteroid belt by the gravity of another passing heavenly body. Had we been perched on that lump of rock 65 million years ago, we would have watched the Earth grow ever larger as we descended upon it. It was not an equatorial aspect that we would have seen but a vision of the planet's nether regions, for the asteroid was approaching from the south. A verdant, glacier-free Australo-Antarctica would have first flashed past. Then a rotating Earth would have displayed a narrowed Atlantic Ocean to the crosshairs of our imaginary asteroidal sights. Above that expanse of water we might have calculated the chances of a splashdown in the sea. Today water covers about 70 per cent of the globe, but 65 million years ago it was even more expansive, covering 75 per cent of the Earth's surface.

As the asteroid came closer its path led it towards the mouth of a great gulf, the entry to the Bearpaw Sea. The equator was behind us now, the asteroid set on a course that, just slightly changed, could so easily have resulted in a near miss. But no, Ground Zero would be squarely in the mouth of that ancient American waterway. Today the site straddles the northern tip of Mexico's Yucatan Peninsula and its surrounding, tropical seas. Near its epicentre lies the town of Chicxulub Puerto, near Progreso, from which the impact site takes its name. When it hit, the rock released the equivalent of 100 million megatons of high explosive, about one hundred times the energy needed to create a global catastrophe, bringing the Mesozoic era—the age of dinosaurs—to a close.[3]

The Chicxulub Crater is lost to the world today. The massive scar, around 180 kilometres in diameter, is buried under a kilometre of limestone rock, deep below a region famous for its white sand beaches and its riot of tropical growth. The fact that the scar has been smothered by limestone, a rocky product of existence itself, and that a diverse albeit somewhat dry tropical forest forms a garnish on its surface, is eloquent of the victory that life has had over this alien death-dealer. But such a thing must have seemed barely possible at the time of impact, for then life came close to being extinguished entirely, at least in North America.

The sea into which the great lump of rock splashed down was about ten metres deep. I imagine it as being rather like the tropical seas of today, with their mixture of azures and luminous greens, and teeming with fish and other marine life. Only back then there were giant reptiles rather than whales and dolphins, and ammonites (like nautiluses) but no reefs of coral like today. Instead, giant clams (some a metre long and shaped like the horns of cattle, others jammed together in enormous colonies) took their place.

Until recently no one knew exactly what kind of extraterrestrial object had collided with the Earth on that day so long ago. Some championed a comet, others a meteorite. But on 19 November 1998, in an article in the journal *Nature*, Professor Frank Kyte of the University of California announced that he had found what might well be a piece of the object itself. It was a fluke discovery—a fragment of corroded rock no larger than a match-head recovered from a core taken kilometres deep in sediments underlying the central Pacific Ocean. The fact that a drill piece just a few centimetres across encountered this tiny extraterrestrial visitor buried deep beneath the ocean gives one pause to reflect upon chance as well as our present mastery of this planet.[4]

The miniature chunk, degraded as it is, still bore the distinctive chemical signature of a carbonaceous chrondrite—a stony meteorite. Such meteorites are rare, constituting just a few per cent of all that fall to Earth but, significantly, they appear to be common in the asteroid belt. The nemesis of the dinosaurs thus appears to have been an errant asteroid, possibly a fragment from a sister planet long since reduced to rubble by some mysterious power, or a fragment left over from the beginning of the solar system that somehow escaped incorporation into a larger heavenly body. Independent studies of the chromium preserved in ancient sediments the same age as that containing the fragment have since reinforced Kyte's findings, for researchers have discovered that they are rich in a kind of chromium similar to that found in carbonaceous chrondrites.[5]

This rock hit the Earth at a speed of at least twenty-five kilometres per second, the atmosphere providing little if any cushioning effect. It almost certainly disintegrated on impact, as stony meteorites are prone to do. Despite this, it opened a hole five kilometres deep in the Earth's crust, blasting thousands of times its original mass into the atmosphere and back into space, where some earthly fragments doubtless became heavenly wanderers themselves.

The great smoking pit left by the impact must have been an incomprehensible sight. How long, I wonder, did the ocean continue to pour into it? So big was the hole that one could not have seen from one side to the other, and so deep that no cliff on Earth today could produce such vertigo. The vast quantities of pulverised rock and blasted life it shot into the atmosphere must have turned the blue planet brown.

Geologists can pinpoint the moment the impact occurred worldwide by the presence of a layer of sediment rich in iridium. Iridium is largely absent from the Earth's crust but is abundant in asteroids and it seems that the disintegration of the carbonaceous chrondrite released enough iridium to act as a global calling card. The iridium layer also contains another indication of disaster—grains of sand known as 'shocked quartz'. The name comes from changes in the crystal structure of quartz produced when intense shock waves pass through it. It develops striations that travel in more than one direction and can even recrystallise into new minerals as a result of the intense pressure of the shock. Alongside these 'shocked' sand grains are tiny glassy spheres, pieces of the Earth's crust that have been thrown high into the atmosphere, then melted to a glassy smoothness as they fell to ground.

This was an impact so intense it could send portions of the Earth spinning into space, coat the globe with iridium and, on the micro-scale, change the structure of grains of sand. How could such awesome forces have been generated, and what were the implications for Ground Zero America? If the location of the strike was unfortunate for North America, the angle of the blow was nothing short of catastrophic. Seismic studies of the crater indicate that the asteroid made a low swipe from the southeast that—like a golfer's chip shot—generated tens to hundreds of times more heat than a vertical strike would have done. It literally fried America, the heat generated by the impact reaching 1000 times that provided by the sun. And the result would have been particularly severe within 7000 kilometres of the impact point, an area that effectively covers the continent.[6]

The atmospheric shock wave must have flattened trees all over North America, creating great piles of timber. There is convincing evidence that the Earth's atmosphere was then about 10 per cent richer in oxygen than it is today. With oxygen levels at about 23 per cent this was a flammable world—when oxygen comprises 24 per cent of the atmosphere, even damp timber will burn. Not surprisingly, evidence has been discovered of vast forest fires following the impact, and a global soot layer has been identified. It appears that much of the northern hemisphere was carbonised by the impact.[7]

The celestial chip shot had other unfortunate consequences, for it took a huge divot of molten rock and debris and propelled it straight up the Bearpaw seaway. Evidence of the result is found in the distinctive sediments that formed as a result of the catastrophe. These rocks differ from similar-aged deposits elsewhere on Earth in that they have two distinct layers. The lower one is a chaotic mixture of materials and may represent part of the 'divot' dumped unceremoniously by the shock wave. The upper layer is better sorted and has thinner layers within it, as if it were laid down by water. Such rocks have been found all along the southern margin of North America, for example on the Brazos River, Texas and near Braggs, Alabama, as well as the Caribbean and Mexico. The exact nature of the upper layer is still debated, but many researchers postulate that they resulted from titanic tsunamis that raked the land a few hours after the asteroid hit.[8]

Scientists think that the sea along the southern coast of North America withdrew as it poured into the crater, only to return a few hours later as an enormous tsunami, or more likely a series of them. Were the waves a kilometre high 'the largest in the history of the world' as some claim?[9] Whatever the case, I cannot imagine them as being anything but titanic, devastating all that remained standing in interior North America.

No humans have ever experienced anything comparable to the asteroid impact, but once—just once in written history—a faint echo of the great impact was directly experienced by a small section of humanity. On 27 August 1883 the island of Krakatoa, lying in the strait between Sumatra and Java, blew apart, releasing as much energy as 10,000 Hiroshima-sized bombs. A vast cloud of pulverised rock blocked out the sun for fifty-six hours. Villagers reported that the darkness was more profound than that of the blackest night, for it was palpable, like a blanket, and it cut off not only sight, but sound as well. In the terrible silence that reigned, they did not hear the explosion that 3000 kilometres away in central Australia was mistaken for the report of distant cannon. They did not even hear the tsunami, as tall as a seven-storey building and travelling as fast as a train, which was to obliterate their world. Thirty-six thousand people met death in that dark, silent instant.[10]

Given the stupendous power of the many violent forces released by the impact of the asteroid it is difficult to imagine any life, except perhaps seeds and microscopic species, surviving in the more exposed parts of North America. Certainly the forests were devastated. The emergent conifers were all destroyed, never to return to the continent, and along with them went nearly 80 per cent of the flowering plant species, including, it seems, almost all of the trees. The destruction off the southern coast was similarly extreme, with even such hardy and uncomplicated creatures as shellfish suffering massive extinctions. Indeed the effect was so profound that even three million years later the fifty-eight species of mollusc then living were but a shadow of their pre-impact diversity.[11]

Further north, in Canada's Alberta province, the evidence left in the rocks is less suggestive of catastrophe, as befits its location near the Arctic Circle of the day. On a cold June afternoon I went to examine the boundary as it is exposed in the badlands along Alberta's Red Deer River. With horizontal rain blowing in my face I trudged through a field of newly sown wheat towards the edge of a precipice. The wind blew like the exhaust from a jet engine as it rose over the rim of the cliff, and I had to force myself to look down at the crumbling unstable-looking slope that dropped away into the stream below. Some years earlier a team from the Royal Tyrrell Museum in Drumheller had discovered a near-complete skeleton of a tyrannosaurus at the base of the cliff. The neck unfortunately disappeared into the rock face and just as they reached the spot where the skull should lie the situation became too perilous even for this resolute and experienced team of excavators. They decided to quit for the season and try to figure out some way to recover the all-important skull. Soon after the entire bluff collapsed, irretrievably burying the skull. The tyrannosaurus had died before Chicxulub and was entombed in a layer around seven metres below that marking the asteroid impact that I had come to see.

As I crossed the slope in search of the boundary layer the ground crumbled under my feet, sending small landslides cascading off the abrupt drop. After some minutes edging my way along I reached the site. The impact horizon was marked by a seam of whitish clay just two centimetres thick. Below it was a narrow band of coal, then the sands of the Alberta badlands that stretched to the level of the tyrannosaurus tomb, and way below that.

As my legs straddled this evidence of catastrophe I wondered at the layers below the clay. The now-vanished creatures entombed there lived in ignorance of an event that would, like an eraser, put an end to their world. What was it like to stand in what is now Canada as the dust that composed the clay layer rained down out of the sky, coating an entire landscape in ghostly, pale grey? Perhaps it was the great breakaway yawning below me, or the dismal weather, but a distinct feeling of the transience of life stole over me as I peered at that thin line that separated two very different worlds.

North America bore the brunt of the impact because it lay in the direct line of the chip shot. Elsewhere, the immediate impact effects were lessened, though the asteroid would unleash a second generation of consequences that would be global in their effects, and which would last not minutes or hours, but months.

Despite all its attendant drama, the direct effects of the asteroid collision cannot account for the massive global extinctions that finished the age of dinosaurs. For that, one must poison the atmosphere. Just how that happened and how long it lasted are issues of intense debate. Some researchers suggest that a vast amount of sulphur entered the atmosphere from the Earth's crust under the impact site. For a few weeks or months this may have created acid rain, particularly in the northern hemisphere. A number of scientists even speculate that the rain acidified the top 100 metres of the world's oceans.[12]

More likely is an idea preferred by Carl Sagan, who coined the term 'impact winter' to explain it. Sagan suggested that the ejecta from the impact made the atmosphere opaque, preventing energy from reaching the Earth's surface. Just how much of the sun's energy was lost is difficult to estimate, but a loss of one fifth for a decade seems reasonable. This would have produced ten years of freezing or near-freezing temperatures across the globe. Even if it did not become that cold, the rock ejected into the atmosphere may have created a twilight sufficiently dim to prevent photosynthesis. In effect, the whole world may have been plunged into a long polar night, thereby starving its plants.[13]

While the immediate effects of the Chicxulub impact on North America are widely accepted by the scientific community, there is less agreement upon the nature of these global, medium-term consequences and how they affected the dinosaurs and other creatures. Recent research in North America points to a sudden extinction of the eight families and fourteen genera of dinosaurs, including such favourites as triceratops and tyrannosaurus, at about the time of the impact, rather than the gradual decline over millions of years that many earlier researchers believed had occurred. Despite such findings, some scientists still maintain that the asteroid had nothing to do with the extinction of the dinosaurs. Nevertheless, the reality of the asteroid impact is now securely established, and I believe that its medium-term consequences best explain the global extinction of the dinosaurs and many other living things.[14]

The great divide in this impact-dominated world was the equator. To its north the consequences of Chicxulub were extreme and palpable, while southward ramifications appear not to have occurred, or at least were much attenuated there. The result is that species which vanished from the north survived in southern lands, for the south suffered no tsunami, no huge shock wave and, perhaps, no fires. This difference was to have ongoing repercussions for life on the various continents. Indeed it can still be discerned in their present faunas and floras, for vegetable dinosaurs still haunt southern forests.

Seymour Island, lying off the Antarctic Peninsula, collected sediment right through the time of the great catastrophe, but these sediments tell a very different story from those of North America. Before the asteroid impact the dominant tree in this Antarctic forest was a relative of Tasmania's Huon pine (*Lagarostrobos*). Looking at the pollen and leaf remains of these trees in the sediments, you would never guess that an asteroid struck the planet at the time they grew. For they, along with the other dominant plants of these southern forests—relatives of the king billy, celery top and other southern pines, the bright-flowered members of the Protea family such as mountain rocket, and the stately southern beeches—sailed right through the crisis unaffected. You can still see their descendants living today, growing in communities that the dinosaurs would have recognised, in parts of Tasmania's south-west.[15]

Anyone fortunate enough to travel to Tasmania will find that a great tradition of timber-working has developed among the inhabitants of that isolated isle. Breathtaking works of art, forged from a timber the colour of rich butter, across which play lustrous rays of light that seem to penetrate deep into the wood, are to be found in the workshops of many small towns. This luminous timber is Huon pine. Count the densely packed growth rings and you'll find that the backrest of a chair took 500 years to grow broad enough to support you, and that a table-top is the gift of 2000 years of history. Such trees cannot be planted for harvest, except perhaps for a people as distant from us as we are from King Canute or Julius Caesar, and they are now so rare, growing in just a few remote valleys, that the living trees can no longer be cut. The wood you see before you is from a tree that grew before Columbus—maybe even before the pyramids.

The furniture is there because Huon pine does not rot. A log will sit in a cold Tasmanian swamp for 10,000 years and, when you pull it out and mill it, will still exude that inexpressible odour that this one tree possesses. Yet Huon pine is even more miraculous, for it survived the Chicxulub impact while its northern relatives perished. It is thus a living vegetable dinosaur, making a last stand in the wettest part of one of the world's most distant islands.

Other vegetable dinosaurs have survived throughout much of Australasia. The rainforests of New Zealand and New Caledonia are still overtopped by magnificent kauri and araucaria pines, trees that once shaded and possibly fed America's dinosaurs. New Caledonia has no fewer than twelve species of *Araucaria* and eight of *Agathis*, in all about half the world's total. All grow as emergents, below which thrive communities of broadleaved flowering plants of an ancient hue, just as they did in North America during the age of dinosaurs. It is difficult not to overstate just how much these southern forests are direct survivors of the age of reptiles.

Australia is the only place on Earth where all three genera of those magnificent emergent araucarians still grow today: the araucarias themselves, those great columnar trees with their impossibly symmetrical growth patterns, the magnificent squat kauri, their bark twisted as if covering sinews and, the most remarkable of all, *Wollemia nobilis.*

When *Wollemia* was found in 1994 the discovery was hailed as being every bit as remarkable as discovering a small living dinosaur. A highly distinctive tree growing forty metres tall, just forty individuals were found growing in a deep canyon about one hundred kilometres from Sydney—a city of four million people. They are the most elegant of trees, pencil thin with bark that looks like Coco Pops, and their male cones actually ejaculate great clouds of pollen into the still canyon air. I was in New Orleans when the discovery was announced. 'Jurassic Bark!' the *Picayune Tribune* proclaimed on page one. I opened the paper expecting to read of a discovery in distant central Asia or Africa, only to find that it had been made an hour's drive from my house in Sydney!

Even if the southern hemisphere evaded tsunamis, fires and other abrupt and fatal catastrophes, something malevolent clearly did happen there. A few species of flowering plants vanished, and of course all the dinosaurs were plucked from the forest, their likes never to be seen again. The dinosaurs, though, were not the only creatures to suffer, for some researchers have suggested that the asteroid left the land bereft of every living creature weighing more than 25 kilograms at sexual maturity and certainly the majority of survivors were far smaller—rat- or lizard-sized.

By sweeping away almost all of the larger organisms the asteroid had reset the evolutionary clock, allowing a few humble if not meek survivors to inherit the Earth. The mammals, which had evolved at the same time as the dinosaurs, had until now played the underdog, for despite more than 100 million years of evolutionary change they had been unable to evolve into creatures larger than a domestic cat. Now all that would change, for the age of mammals—the Cenozoic era—was at hand.

THE REORDERING OF
NORTH AMERICA

Both before and after the asteroid impact, capacious basins were accumulating sediments in what are now the states of New Mexico and Colorado. For decades the coal-rich sediments have attracted miners, who obligingly have stripped bare successive layers of rock, laying open the book of life. The extraordinary tale of destruction and redemption written in carbon and preserved on pages of stone has amazed the palaeontologists who have read it.

In the sediments below the iridium layer fossilised leaves, fruit, trunks and pollen of an ancient forest are preserved. Palms, relatives of the nutmeg and the plane tree, the alarmingly symmetrical Norfolk Island pine, and massive cedars related to the alerce (*Fitzroya*), Chile's famed national tree, can be found, along with many other species. The fossils bespeak a riot of diverse growth. But a very different story is revealed in the metre of sediment immediately above the impact zone. Gone are the trees. Gone is the forest. Instead there is nothing but apparently lifeless clay, eroded from slopes that had been stripped of their stabilising plant cover. I find it difficult to imagine North America as a vast field of churned rock and clay. From north to south—all brown, little moving. It is an alien and perhaps unimaginable concept. Yet here we find testimony of it.

If you look at the clay through a high-powered microscope you see that the devastation was not quite complete. Minute signs of vitality come into view; tiny fern spores are reminders of the gentle yet irresistible force of procreation, and of the indomitable nature of life.

It took centuries before the brown of North America once again turned to green. The species that won back the continent was not one of the advanced flowering plants nor even a pine but, as our microscopic examinations indicate, a humble and ancient fern. Its spores reveal that it was an adventurous, colonising relative of the Malaysian climbing fern (*Stenochlaena*), a beautiful plant that today is restricted to the Old World tropics and the greenhouses of avid gardeners who grow it for its graceful form. This solitary pioneer would slowly win back the blasted ground, its graceful fronds shading the surface, its stolons binding the soil. The story of its conquest is told in the first few centimetres above the impact layer. There, somewhere between 70 and 100 per cent of all fern spores, which constitute virtually all the evidence of life in the rocks, belong to this one survivor of the diverse, pre-impact vegetable world.[1]

A similar drama of devastation and recovery was played out, albeit on a much smaller scale, after the 1883 eruption of Krakatoa. Among the first living things to return to the sterilised pile of rock were eleven species of ferns, including our hero of the asteroid impact recovery, the Malaysian climbing fern. This fern dominated away from the coast, and by the early 1930s had grown so dense that scientific parties investigating the island could make little headway through the thickets and often got lost.[2]

Higher in the first metre of sediment preserved in the New Mexican coal mines the diversity of fern spores increases, indicating that other ferns had joined *Stenochlaena* in colonising the land. A lucky palaeontologist might come upon a fragment of a leaf, indicating that some herbs and palms had also survived the catastrophe, perhaps in the lee of a mountain range or in the far north. Their absence from lower levels suggests that it took these survivors thousands of years to migrate over the devastated landscape to New Mexico and that they represent species very different from anything that lived on the site before.

Evidence suggests that massive plant extinctions occurred as far north as what is now North Dakota, where 80 per cent of all species vanished.[3] Such extinctions might have occurred even within the Arctic Circle in North America, but with one exception evidence of such destruction is absent on other continents. The exception comes from Hokkaido, the northernmost island of Japan, where an examination of post-impact sediments reveals, as in North America, nothing but fern spores. East Asia, too, it seems, was in a direct line to receive the worst of the impact and, even then, this region via Beringia was North America's principal conduit to the rest of the world. This devastation must have slowed any immigration into the continent, if it were at all possible at that time.

In the sediments in the next metre above the impact zone in New Mexico you no longer need a microscope to find evidence of life, for fragments of leaves are more common. But these are special leaves and they speak to a botanist of an endless story of succession. After any disturbance there are certain plants that will be first to set root in the bared soil, and they in turn are followed by other species in what is known as a plant succession. The first to arrive are called pioneer plants, and these Daniel Boones of the vegetable world have typical and readily recognised growth habits.

The fossil leaf fragments from New Mexico inform us that relatives of the beefsteak plant (*Acalypha*) and poinsettia (*Euphorbia*) were among the first pioneers. These ancient plants, however, did not bear much resemblance to their living relatives for they had long narrow leaves, rather like those of a willow. These leaves are characteristic of plants growing along watercourses and their shape offers minimal resistance to flowing water. As a result, the leaves tend not to be pulled off during floods and the force of water fails to uproot the plant. Such leaves evolved with the disturbance that torrents create, and the plants that bear them live in locations such as floodplains and are especially adapted to colonising newly bared ground.

What is extraordinary about the ancient leaf fossils of New Mexico's coal mines is that they are not preserved in the distinctive sediments laid down on floodplains or beside streams, but in coal-beds. Coal forms in swamps and, in a normal world, streamside vegetation has no business growing in such quiet backwaters. This anomaly tells us a remarkable thing: the streamside species had escaped their normal habitat restrictions and had colonised an entire continent. They were not that well suited to growing in swamps or other still areas, but in the absence of competition could exist there. The key to their successful spread was their ability to survive disturbance and to colonise naked earth—a quality developed during millions of years of natural selection on countless newly stripped watercourses and floodplains. How long these streamside pioneers dominated the New World is unknown, but to judge from the various soil horizons that formed and were then buried (a process which takes thousands of years, for soil forms but slowly), their reign lasted many, many millennia.

In the sediments extending from this zone to 200 metres above the impact layer we see a world slowly beginning to conform again to the eternal rules of ecology that keep things in their place, and which bring order and meaning to our lives. The narrow-leafed plants once again become associated with streamside sediments, indicating that they have been put back in their place by competitors. The plants that form the coal once again have leaf shapes characteristic of species that grow in undisturbed forests and swamps.

None of these plants, however, have close relatives in the pre-impact rocks. Almost all have come from somewhere else to take their place in this reassembled world—but from where? The origin of many of these plants becomes clear when we examine the rocks lying between 270 and 420 metres above the impact layer. There we find relatives of the magnolia (Magnoliaceae—ancient flowering plants), and basswood (Tiliaceae), along with the Chinese water pine (*Glyptostrobus,* a relative of the sequoia). What is remarkable about all of these trees is that they were, as their living relatives largely still are, deciduous. Here is an astonishing anomaly—a tropical continent dominated by deciduous trees, which even grew along its southern coast in what is now New Mexico! This second breach of the rules of ecology provides a vital clue to where the post-catastrophe colonists came from. Such trees grew in only one area before the asteroid impact—around the North Pole.

To understand how this bizarre circumstance arose we must imagine North America divided into a great series of ellipses with Chicxulub as the point of origin. The nearer the impact point, the more likely the devastation was to be total. Seen this way, just three locations were sufficiently remote or sheltered to act as a refuge to plants—in the lee of the mountain ranges of the Sierra Nevada and Appalachians, and far to the north, inside the Arctic Circle. We can only speculate about what might have survived in the lee of the mountains, but much more is known about survival at the poles. It should also be noted that 80 million years ago North America was located 10 degrees further north than it is today. In one northern hemisphere group after another, whether in the depths of the Arctic Ocean (as plankton, snails or clams), or in the forests on the northern lands, the Arctic Circle was the only sure sanctuary from the devastation of Chicxulub.[4]

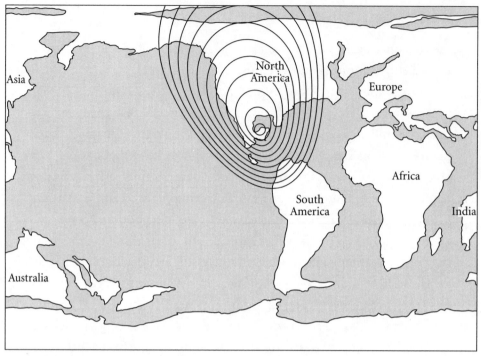

Ground zero America: ellipses radiating from Chicxulub show the asteroid's zone of greatest destruction 65 million years ago.

During the age of the dinosaurs a distinctive and peculiar kind of forest grew in that part of North America close to the pole. As elsewhere on the continent this flora was a combination of conifers and smaller flowering plant species, but the dominant conifer in this forest—which extended from the Colville River, Alaska (then located at 85 degrees north) to southern Alberta—was a deciduous species closely related to the dawn redwood (*Metasequoia*). These forests could survive in the far north because, during the age of dinosaurs and for millions of years thereafter, the Earth was in full 'greenhouse' mode. There was no ice at the poles, and even though these regions experienced up to six months of darkness per year the climate there was relatively mild. At what was about 85 degrees north, on the north slope of Alaska, the mean annual air temperature was 2 to 8 degrees Celsius, comparable to temperatures in the Pacific Northwest today.[5]

Imagine a tree growing in such circumstances. Throughout the summer the sun would slowly circle overhead, and it would disappear below the horizon during the long polar night. Where would the tree direct its leaves to collect the precious energy of the sun's rays? Sunlight critical for its growth would be coming from the sides as well as from above. The optimal shape for the tree to collect this polar light would be a cone, standing alone, with no neighbours within shadow range to steal the precious lateral light. The dawn redwood grows in a rather conical shape even today. It is unclear how its ancestors might have excluded competitors from shading their light-collecting cone. Perhaps they did so underground, the subterranean halo of their roots forming a *cordon sanitaire* around them. Whatever the case it seems that all plants in this unusual forest—even the conifers—were deciduous, yet few species bore the deeply lobed leaves so typical of deciduous forests today. Instead, leaves were often large and spear- or oval-shaped. Evidence of vines is all but absent, and it appears likely that any undergrowth was sparse. Despite being so maladapted to life at lower latitudes, this swath of deciduous trees was to contribute disproportionately to the recovery of North America's vegetation and to remain important in the continent's forests for ten million years after the catastrophe that devastated the evergreens.[6]

Some other survivors probably did not come from the polar refuge, among them possibly the ancestors of North America's few endemic plant families. Today the continent boasts just four families found nowhere else: the Crossosomataceae, which are a small group of shrubs found growing on difficult soils derived from oceanic crust, the geranium-like Limnanthaceae, the lonely Simmondsiaceae (represented by a single species) and the boisterous Garryaceae, a small group best known by the boojum tree of Dr Seuss fame. All of these ancient families are restricted to, or have distributions centred on, the south-west of the continent. They were presumably in existence long before the Chicxulub impact, and perhaps found refuge from flood, fire and asteroid in the lee of the Sierras.

Other species may also have survived in similar places, for even during the age of the dinosaurs small deciduous plants grew along the banks of streams and on disturbed ground at low latitudes. It was in these habitats that the flowering plants first evolved as small flimsy things (we would think of them as bushes) growing where, in earlier ages, only ferns had survived. The need to shed leaves at the poles is obvious—leaves are nothing but targets for herbivores during the long polar night. But for plants growing by streams at low latitudes the need to shed leaves is less apparent. We must remember that these were small plants, not trees. As the soil dried out during summer they had to conserve moisture. Dropping their leaves and going to sleep seems like a sensible adaptation in such circumstances. If such species were growing in the lee of America's mountains they may well have survived the catastrophe and gone on to recolonise the barren earth.

During times of stress, deciduous plants have one decided advantage over their evergreen neighbours: they can 'shut down', even when winter is not on its way. This very convenient response can be seen today when caterpillar plagues denude an oak in full flush. Despite the loss of its new leaves the tree can resprout after the caterpillars have gone. This ability to switch off during adverse times would have given the deciduous trees a valuable advantage if the asteroid had struck in the fall—the trees would have already withdrawn their nutrients from their leaves and begun preparations for their winter sleep. Under such circumstances we can imagine them as pre-adapted to surviving a nuclear winter of six months duration or less. If, however, the asteroid struck in spring the trees might have been caught in full flush. Six months of nuclear winter would then have deprived them of a whole summer, leading to eighteen months without nutrition. Despite their adaptations, surely this would have killed them as efficiently as it killed their evergreen neighbours?

Most of the plants of the ancient North American forests are now extinct on that continent, driven from it by violent swings of climate over the last 65 million years, although several species found a last refuge in Asia. The deciduous katsura tree (*Cercidiphyllum*) of the Far East, the charming *Trochodendron* of Japan and Taiwan, the ancient ginkgo (which survived in North America until five million years ago), and even the dawn redwood (*Metasequoia*) fall into this category. In fact, because of the dawn redwood's dominance within the Arctic Circle, it is as important to the history of the northern forests as the Huon pine is to the Antipodes.

The dawn redwood grew in North America from the age of the dinosaurs until about eight million years ago, when it made a last stand in the swamps of what is now eastern Washington state. Although well known as fossils, they were believed to be extinct until a small population—just 1200 trees—was located in a steep isolated valley in Hubei Province, south-central China, in 1941. They were discovered as a result of one of the most vicious wars ever fought—the Sino-Japanese conflict of 1937–45. A series of disastrous defeats forced the Chinese government to retreat into the mountain fastnesses in its west. There it began to assess its resources. Among the inventories carried out was one of timber, during which a sharp-eyed botanist noted the unusual trees. Ironically, in the same year that the trees were discovered on the Chinese side of the lines, on the other side of the battle front a Japanese palaeontologist was naming some new fossil plants from Japan. He called the strange, deciduous conifer leaf fossils *Metasequoia*, giving the tree its own distinctive scientific (generic) name for the first time.[7]

In 1946, pioneering collaborative work between Dr Elmer Merrill of Boston's Arnold Arboretum and the Chinese government brought the first living dawn redwoods out of China. The trees must have appreciated being returned to their ancient homeland because by 1952 some planted in Portland, Oregon, had developed cones—the first to be seen in North America for eight million years.[8]

To return to the issue of how North America was reordered, we need to address the question of whether only indigenous species contributed to the recovery, or if North America was open to immigration from elsewhere. This is a difficult question to answer, but some general inferences can be made.

North America has not remained entirely immobile over the past 65 million years, for it has drifted south and rotated somewhat westward. Sixty-five million years ago the North Pole lay squarely over northern Beringia, the now-submerged isthmus that links Asia with North America. Even in a greenhouse world this would have made immigration from Asia difficult, for few species would have been able to survive in those strange polar forests. Compounded by the devastation the asteroid wrought in north-east Asia, any chance of successful migration from Asia, at least soon after the impact, seems slight.

Europe at that time was an island archipelago. The English area of this ancient island realm possessed a broad connection with North America via Greenland and the Faroes, while Scandinavia was connected via Greenland. Both bridges, however, lay over the Arctic, so they were of limited use to life, even given the rather warm global climates of the time.

While considering continental drift, it is worth placing North America relative to South America during the period when life on the continent was slowly beginning to reorder itself. This is important, for proximity affects the likelihood of dispersal across the sea. South America had been drifting towards North America for 20 million years before the asteroid hit. Indeed, it probably made land contact just before Ground Zero. After this brief continental kiss South America began to waltz—at about eight millimetres per year—to the south-west, out into the Pacific. Whatever connection had earlier existed between the two continents was now lost. North America faced only sea to the south, and so it would remain for 63 million years.[9]

The apparent isolation of North America in the few million years after the asteroid impact stands in stark contrast to what had gone before. While scientists are still uncertain about what connections existed between the continents late in the age of dinosaurs, there is ample evidence from flora and fauna to indicate that they must have been substantial. The flowering plants—which rose to dominance in the last few million years before the event at Chicxulub—rapidly achieved a worldwide distribution, as did many kinds of dinosaur, as well as advanced mammals such as placentals and marsupials. The global distribution of these organisms is a persuasive argument that contact between the continents was extensive.

Throughout this time North America continued to remodel itself geologically. Prior to the great catastrophe, America's west had gone into paroxysms as forces deep within the Earth's crust began pushing vast masses of basement rock skyward in a dramatic episode of mountain-building known as the Laramide Orogeny. The Bearpaw Sea that had divided the continent drained away as uplift raised the land's surface until it had shrunk to a brackish expanse known as the Cannon-ball Sea, so-called because of the strange, rounded concretions that formed in its sediments. Sixty million years ago, when the Cannonball retreated for the last time from Montana and North Dakota, North America, east and west, was finally united across a broad front by this awesome orogenic process.

While ructions were afoot in the west, the eastern side of North America experienced sedate continuity. With the exception of ice age changes in the north, the ancient Appalachians would have taken on the basic shape they possess today. The eastern coastline continued to sink into the sea at an almost imperceptible rate, accumulating sediments that in eons to come would help scientists to track the rise and fall of sea levels worldwide. This slow sinking has also given us the swampy coastline that extends from Florida to southern New England. The Sea Islands of the south-east coast and the magnificent barrier islands of Cape Hatteras in North Carolina would all have been impossible on a more active continental margin. The coast to the north would have been much the same had it not been for the ice age, which depressed and gouged it, only to see it rise when the ice melted, giving the rocky indented coast typical of Maine today.

This then is the geographic field in which the drama of returning life unfolded. North America's only connections to other land after the asteroid impact were polar portals providing limited potential contacts with Europe and Asia. Consequently, for the first few million years after Chicxulub, plant evolution on the continent seems to have continued on a uniquely North American path, with few immigrations and emigrations.

North America's recovery from the impact was to remain slow. It would take millions of years for the few surviving plant species to diversify. Even by 60 million years ago, five million years after the catastrophe, plant diversity had increased only a little—forty to fifty species were present in areas such as New Mexico, a pathetic total compared with the seventy-five or more that grew there before the impact, and it would be ten million years before the deciduous species gave way to the evergreens.[10]

Trees evolve far more slowly than other organisms such as mammals. One need only consider the rat-like creatures living 65 million years ago that were to morph through time and space into elephants and whales, yet many of the trees growing during the last part of the age of dinosaurs would not look out of place in the modern world. The bogs and lakes of the time preserved leaves easily mistaken for the contemporary plane, beech and redwood. Indeed, some of these tree lineages have hardly changed through 65 million years of Earth history.

The asteroid impact dictated which trees henceforth would grow in

America, and its effect can still be seen in the foliage that graces the continent today. The same is true for all groups with limited abilities to migrate, and none were more profoundly affected than the reptiles and amphibians.

THREE

OF HELLBENDERS AND
HOOFED CREATURES

The rocks of the United States of America have sequestered away at least one rich fossil field for almost every million years that have passed since Chicxulub. Canada has not been so fortunate, for the ice age bulldozed away most of the rocks that must have accumulated there over the last 60 million years. In Alberta some rocks of this age do remain, lying atop the dinosaur-bearing sediments, but they often support dense vegetation and are thus poorly exposed. Only in southern Saskatchewan are better examples to be found, and here the record largely repeats that found further south. The fossil record of Mexico promises to be rich, though as yet it has not been studied in great detail. The story of this book will therefore be told mostly through the rocks of the United States.

North America's rich treasure-trove of fossils has spawned a formidable science—that of contemporary vertebrate palaeontology. Europe is where the study of fossils first developed, but the completeness of the

North American record has allowed its researchers to remodel the science according to their particular needs and challenges, so that today it is something quite different. The US now trains and employs more palaeontologists than any other nation and its innovative techniques, explorations and studies are the global engine of the discipline. Hence, its inhabitants know how life unfolded on their continent in great detail. North America, paradoxically, is also the global centre of Creationism, whose dogmatic followers believe that the Earth was formed just 6000 years ago. Enigmas such as this abound on the continent, and indeed seem typical of it.

The rocks of Wyoming, the Dakotas and Montana provide a comprehensive record of the 13 million years following the extinction of the dinosaurs. Deep faulting and folding created basins between the peaks of the Rocky Mountains as they thrust skyward between 65 and 55 million years ago, and it is from within these basins—some of which collected over 11,000 metres of sediments—that the fossils making this tale possible were preserved. The story of animal survival and evolution as told in these rocks is a surprising one, for the survivors of Chicxulub were a motley lot with, at first glance, little in common.[1]

Despite the dismal fate of the dinosaurs, many other reptile groups sailed through the extinction event as if it were a picnic, and today their descendants comprise some of North America's most distinctive and venerable animal inhabitants. The turtles and terrapins survived so well that the period immediately following the catastrophe should perhaps be named after them. As testimony to this age of turtles, the remains of no less than nineteen species have been found in sediments lying above the impact layer in north-eastern Montana. Sixteen were local survivors of the impact and the other three species were immigrants to the area. This constitutes the most diverse assemblage of turtles ever to have existed, but it is the abundance of their remains that makes them a worthy namesake for the age. Their fossilised bones are everywhere, indicating that they dominated many habitats. Remarkably, survivors from this American age of turtles are still with us, and what extraordinary creatures they are.[2]

I shall never forget my first meeting with an alligator snapping turtle (*Macroclemys temmincki*). It was at an aquarium in Louisiana, and thankfully I was on the other side of a pane of glass. The mud-coloured, warty creature, evil of eye and fearsome of beak, with a long scaly tail like a crocodile and serrations running down its shell, seemed to me to be the Buick of the turtle world. It sat in its dim tank open-mouthed, beckoning entry to passing fish with an obscene pink, worm-like object on its tongue whose restless movement was the antithesis of the motionless body of the creature. This hypnotic excrescence is used to lure animals within snapping range, and it did a marvellous job of mesmerising me.

At up to 110 kilograms the alligator snapper is the largest freshwater turtle in the world. It is a North American production in a way that few other large creatures are, for its lineage evolved on the continent some 100 million years ago and it has since survived in suitable habitats throughout the land. Its only living relative, the snapping turtle (*Chelydra serpentina*), even extends into southern Canada. The family lived only in North America until three million years ago—when the snapping turtle crossed into northern South America.

The soft-shelled turtles, like the colourful and familiar eastern spiny softshell (which even survives in grossly polluted rivers), and the pond turtles (such as painted turtles, sliders and cooters), are also survivors from this time. Unlike the snappers, however, they are not unique to North America and are found throughout the tropical regions of the world as well as sometimes in more temperate latitudes. Other turtle families, such as the Dermatemydidae, have been less successful. In the age of turtles they were distributed over much of North America, but today the sole surviving member of the family, the Central American river turtle, finds its last redoubt in the rivers and streams of Mexico and northern Central America, which it never leaves except to lay eggs.

The turtles shared the rivers and ponds of their post-impact world with a group of now extinct crocodile-like creatures known as champsosaurs. Remarkably, for a time they even seem to have benefited from the asteroid crash, for the fossilised vertebrae of champsosaurs found above the impact zone in north-eastern Montana are nearly a centimetre broader than those found below it. Their fossils make up as much as 35 per cent of the number of individuals recovered from some post-impact sites, suggesting a prodigious abundance. Perhaps they ate turtles as well as fish.[3]

Several kinds of crocodiles as well as at least four species of alligator also survived, but one alligator known as *Brachychampsa* seems to have been eliminated by the asteroid. Just why this particular species, alone among alligators, was selected for asteroid-induced extinction is unknown.[4]

In stark contrast to the remarkable survival of the turtles and crocodiles, most North American lizards suffered a dismal fate. The remains of what appears to be an ancient relative of the Gila monster have been found below the iridium layer, as have the fossilised bones of primitive goanna-like species. The asteroid drove them from the continent and it would be millions of years before such creatures would stalk the terrain of North America again. Before the asteroid struck, at least five species of whiptails and race runners (family Teiidae) inhabited what is now Montana and Wyoming. All vanished at impact. The sixteen species of whiptails and race runners inhabiting the US today are all descendants of later immigrants from South America, where the remainder of the 230 living species in the family can be found. Today the knob-tailed lizards (*Xenosaurus*) are found only in Central America and along streams in southern China. The one species inhabiting Montana at the time of the asteroid survived the impact, as did a single kind of skink.

The alligator and glass lizards (family Anguidae) actually prospered after the impact, becoming the most abundant of all lizards. These curious creatures are covered in a heavy armour so unyielding that were it not for a groove which runs down their sides (allowing a break in their bony defences), most would be unable to breathe. Many, including the strange Caribbean galliwasps and the glass lizard, are obscure burrowers or ground-dwellers, and are carnivores.[5]

The remains of just one snake are known from the pre-impact sediments of Montana. It belongs to the family of pythons and boas (the Boidae), a group that first evolved just prior to Chicxulub. Unfortunately the solitary Montanan specimen, which is possibly the oldest python in the world, was an arthritis sufferer, and its bones are so deformed by the disease that it is difficult to identify it with certainty.[6]

The amphibians are an even more archaic group than the reptiles, yet they seem to have been very successful in avoiding extirpation by asteroid. Their ancient, fossilised bones inform us that creatures akin to the hellbender (family Cryptobranchidae), which at a metre long is the largest North American amphibian, and the mudpuppies (family Proteidae) were among the survivors. Both have relatives living elsewhere: east Asia for the hellbender, caves in the Balkans for the mudpuppies. These creatures are happiest when hidden under rocks in rivers and streams. They are occasionally disturbed by fishermen searching for bait who, aghast at the hideous appearance and large size of the 'devil dog' (aka the hellbender) or the touch of the mudpuppy, invent tall stories about these amphibians, accusing them of coating fishing lines in slime, driving away game fish and having a poisonous bite, none of which is true.

Among the other amphibian survivors were the amphiumas. They evolved in North America as one of a handful of vertebrate families unique to the continent. These eel-like amphibians can exceed a metre in length and are secretive residents at the bottoms of ponds, rivers and bayous. Their status as ancient native Americans has not spared them from the wrath of fishermen, however, who 'detest them as a nuisance preying on virtually everything that swims, including other amphiumas'. To top it off they too are supposed to have a nasty bite.[7]

Included with the surviving frogs were relatives of the bizarre Mexican burrowing toad. This is another secretive creature—an adept burrower that only comes out at night. At first glance it looks like a common red and black toad, but when disturbed it can inflate itself to a ridiculous extent, coming to resemble 'a walking balloon'. Its other physical peculiarities include the absence of a breastbone, pupils like a cat, and a tongue attached at the back of the mouth (identical to us but the reverse of all other frogs whose tongues are hinged at the front). This strange creature is placed in another unique North American family—the only American frog or toad so distinguished. Relatives of the archaic red-bellied toads (family Discoglossidae), which are now found only in the Old World, are also known from North America at this time.[8]

In all, four living families of amphibians (one of toads, three of newts and salamanders) are unique to North America. This high level of endemism is due entirely to the survival of so many amphibians through the asteroid impact. Indeed, it is difficult to find any evidence at all for North American amphibian extinctions at the time. Why should these delicate creatures have emerged unscathed from such an unparalleled disaster?

To understand the nature of the catastrophe at Chicxulub 65 million years ago one needs to explore why some species vanished and others persisted. The clue is to correlate survival with habitat type, for it is clear that many of the survivors were aquatic—the turtles, champsosaurs, crocodiles, hellbenders and fish such as the gars and the ancient paddlefish. It seems nonsensical that aquatic ecosystems should be less affected by the asteroid than either marine or terrestrial ones, but there are good reasons why this might have been so. Both marine and terrestrial ecosystems are entirely reliant upon plants. The oceans are very intolerant of a nuclear winter because phytoplankton is so short-lived. Even a brief interruption, if widespread enough, can cause massive damage. Land-based plants are a bit more robust, for they can 'shut down' for months, resprouting from roots or seeds. Aquatic environments, however, are somewhat independent of plants in the short term, for their chain of life is in part detritus-fed. If photosynthesis were interrupted for six months or so, many aquatic species such as turtles might hardly notice, for dead leaves and other plant matter would continue to be flushed into the rivers with each flood, providing food for the decomposers such as bacteria and the tiny creatures that in turn feed on them and which feed the turtles. Furthermore, water has an enormous capacity to absorb heat, and creatures living in deep pools or potholes in rivers may have been shielded from the intense temperatures of the initial impact; neither forest fires nor shock waves greatly affected them. It is clear, too, that the river-dwellers survived any tsunamis in sufficient numbers to inherit this post-holocaust world. Acid rain, however, would have devastated them, so we can surmise that this was not a major feature of the impact.

That is not to say that the aquatic ecosystems escaped entirely unscathed. All thirty species of the notoriously vulnerable freshwater mussels that were present in pre-impact layers vanished. Indeed, mussels are absent for the first 130 metres of sediment laid down in Wyoming's Fort Union Formation after the impact layer (and they are probably absent elsewhere in post-impact rocks), with a mere four species reappearing after a long absence. Freshwater snails do a little better, with several species sailing across the boundary.[9]

The differential survival of the various land-based species demands another explanation. Among the least affected groups were those that today consist largely of burrowing forms such as the Mexican burrowing toad and the glass lizards; their subterranean habits may have protected them from the worst effects of the initial blast. The groups that suffered most were the highly active animals such as the whiptails and goannas, which live in relatively exposed places thereby leaving them vulnerable to the initial impacts. Today such species are much more severely affected by wildfires than are burrowers and this was probably true in the past as well.

The survival of so many reptile and amphibian groups in North America was to have important implications for the evolving fauna. Because they are cold-blooded, reptiles and amphibians found the polar Bering land bridge inaccessible, ensuring many survivors stayed where they had evolved, and new reptile invaders from Asia were few when compared with the influx of warm-blooded animals. As a consequence it is the reptiles and amphibians, with six strictly endemic families, that represent the heart of endemism in North America today. Over the past 30 million years a cooling climate has driven many of these ancient North Americans south of the Rio Grande, so now one must travel to Mexico to see them. A herpetofauna can be seen there that is among the most distinctive and diverse on Earth—a true gift of apocalypse survival.

It was not the reptiles, however, that were to shape the post-Chicxulub world order, but the mammals. For 100 million years they had been losers in the race of life, rat-sized creatures whose evolutionary opportunities were thwarted by the terrible dinosaurs. How did they cope with fire, flood and atmospheric catastrophe? Were it not for the habits of a hard-working North American ant known as *Pogonomyrmex* this would be a difficult question to answer.

Pogonomyrmex's importance stems from the fact that it likes to get a jump-start in spring. Being cold-blooded, the ants find it difficult to get moving after a hard winter such as occurs in Montana, but they have developed an ingenious solution to this problem. They build crater-shaped rings around the entrance to their underground nest, and these they ornament with anything that can hold the heat of the sun. The teeth of tiny long-extinct mammals, often stained a dark colour during millions of years underground, are just about perfect as heating pads. Thus, with the help of teeth that may have pulverised ants 65 million years ago, *Pogonomyrmex* are often up and about feeding long before other insects have woken from their winter slumber. Over the years a colony can accumulate thousands of fossilised mammal teeth, a fact that was discovered and exploited by American palaeontologists more than a century ago. These teeth tell a fascinating tale about the fortunes of ancient North American mammals.

The fact that an ant could carry the teeth (and sometimes whole jaws) of extinct mammals should give you some idea of the diminutive size of these ancient creatures. Although most mammals were rat- or mouse-sized they were exceptionally diverse, belonging to three distinct orders—the placental mammals, the marsupials (such as the Virginia opossum), and a primitive group called multituberculates, which were possibly distant relatives of Australia's egg-laying mammals, the platypus and echidna. The fate that befell the mammals is a story best told in the sediments of Garfield County, Montana, where palaeontologist David Archibald has spent years combing the outcrops, sifting through tonnes of sediment, searching for the tiny remains.

Earlier studies had suggested that about half of North America's mammal species survived the great catastrophe, but we now know that this is wrong. The error was caused because fossilised remains of different ages were mixed as ancient creeks cut through earlier sediments and redeposited fossils. It now appears that perhaps as little as 7 per cent of Montana's mammal species survived—just three or so species out of nearly forty, and most were primitive multituberculates.[10]

Archibald has found that the rocks below the impact zone are extraordinarily rich in marsupials. Many people think of Australia as the home of the marsupial because well-known species like kangaroos and koalas can be seen there. It seems likely, however, that the marsupials originated in North America or in Eurasia around 120 million years ago. Archibald has identified thirteen species belonging to three separate marsupial families in the Montana sediments, among them *Didelphodon*, a cat-sized and possibly aquatic carnivore and the largest mammal of its day.

As mammal lifeforms marsupials generally dominate in pre-impact rocks in North America, but their numbers were devastated by the collision. In rocks laid down just after the event in Garfield County, only one marsupial species, called *Peradectes pusillus*, is present. This tiny creature (its molar teeth are just two millimetres long) could not have weighed more than fifty grams. It is a member of the same family (the Didelphidae) as the Virginia opossum. With seventy-six living species, the Didelphids dominate the South American marsupial fauna. All of them possess one bizarre characteristic—their sperm swim in pairs. No one knows why they do this, nor even how the lucky pair that reaches the egg decides which will fertilise it. Still, the success of this group in surviving the impact, while the other North American marsupial families perished, says something for their resilient if sometimes bizarre biology.[11]

A curious thing about *Peradectes pusillus* is that it is not present in the pre-catastrophe sediments of Garfield County. Even the genus *Peradectes* is unknown there, indicating that this miniature marsupial was an immigrant—but from where? It is unlikely that *Peradectes* came from Asia, for Asia had lost most of its marsupials long before the asteroid impact, with just a few specialised, ferret-sized predatory types persisting there, and even these were extirpated during the catastrophe.[12]

Peradectes, I suspect, must have survived in North America either at the pole or in one of the refuges created by mountain ranges further south. In any case it is clear that all thirteen of Garfield County's original marsupial species fell victim to the catastrophe, and that the North American marsupials never recovered. The descendants of *Peradectes* dribbled along, a few species finding a home in the reconstituted forests of North America, until the last became extinct following climatic change around 18 million years ago. A marsupial finally returned to the continent 16 million years later, when the ancestor of the Virginia opossum invaded from South America.

The marsupials still thrive on the southern continents of Australia and South America. They reached South America from North America around the time of the asteroid impact, but their early history in Australia is frustratingly unclear, for there is a gap in the mammal fossil record from 120 until 54 million years ago. Still, it seems that their survival in the south is due to the distance from the impact zone.

The squirrel-like multituberculates (their name means 'many bumps' or tubercles in reference to their complex teeth) were the most ancient mammals in North America, having existed for 130 million years when the asteroid struck. Their remains are plentiful in North American rocks and, while they were close to being obliterated, the continuity of two or three species across the impact zone indicates survival somewhere near eastern Montana. These ancient creatures do better than the marsupials, for within a million years eight species are again living in eastern Montana, all resulting from rapid evolution in North America. Multituberculates had been extinct in Asia for some time before the impact, so it is remarkable that within two million years some newly evolved North American species had mounted a successful Asian invasion. Their tenure there and in Europe was destined to be brief, however, for by 50 million years ago they vanished from both continents. North America provided a last refuge until competition from primates, rodents and primitive hoofed mammals eliminated the last multituberculate about 35 million years ago.[13]

The third and final group of mammals whose teeth are preserved in the sediments of Garfield County is our own—the placental mammals. In Garfield County, as in all of North America, remains of placental mammals are comparatively rare in sediments below the impact layer. In the Lance Local fauna (an important collecting horizon in Wyoming), for example, they comprise just 10 to 15 per cent of mammals. About eight species have been recorded in pre-impact rocks in Garfield County, but because most are known from just a few fragments (as little as half a tooth) it can be difficult to identify them precisely and determine how well they fared during the catastrophe.[14]

It seems possible that one of these species survived: the tiny, primitive hoofed mammal *Procerberus formicarum* (its species name means 'of the ants', in tribute to those tireless fossil collectors). It originated in Asia, evolving there tens of millions of years earlier. The remaining nine of the ten species found in post-impact rocks in Garfield County are immigrants that arrived from Asia after the impact. Indeed, researchers are divided about whether *Procerberus* arrived just before or after the asteroid impact. The placentals had become the dominant mammals in central Asia long before the Chicxulub event and they were the only mammals to survive the disaster there. It seems possible that all of North America's post-impact placentals were Asian immigrants.[15]

One tangled history that illustrates the uncertain state of our knowledge of the rare North American placentals concerns our own lineage—the primates. As late as 1979 a single tooth of a primate-like creature called *Purgatorius ceratops* was hailed as the oldest evidence of a primate in the world, and the only one dating from the age of dinosaurs. The tooth was found in the Hell Creek Formation of Montana, but later work determined that these rocks were in fact more recent. To make things worse, no one is sure anymore whether *Purgatorius* or any of its relatives (known as plesiadiforms) were primates at all, for they may be a separate branch related to flying lemurs. Such uncertainties are the purgatory of the palaeontologist wrestling with the study of a solitary tooth, yet an attempt must be made to understand such fragments if the history of life on Earth is to be recovered.

The mammals, being small, would have stood a fair chance of surviving the immediate asteroid impact, but because they are warm-blooded and need to eat every day, they would have been highly vulnerable to medium-term disruption of their food source. Unless some could hibernate (unlikely because of the tropical climate), this would have been a severe disadvantage during a nuclear winter. If medium-term effects such as a nuclear winter are in fact responsible, we should see large-scale extinctions of mammals in the southern as well as the northern hemisphere, for both appear to have suffered such effects. At present the fossil record of the southern continents is too poor to test this idea.

One final group of which we know very little in terms of their survival during the great catastrophe is the birds. Primitive shorebirds, some looking like waders sporting the head of a duck, have been found in North American rocks laid down before the asteroid, and again in rocks some millions of years later, but it is uncertain whether they survived in North America or returned from elsewhere.[16]

If the general pattern holds, however, we can imagine that fearsome mortality occurred among the feathered tribes in North America, for they are mostly non-burrowing, warm-blooded and do not hibernate. Indeed, it seems possible that the southern hemisphere was a critical refuge for this group, but this will not be known until the fossil record yields more avian secrets.

Following the impact it took about 400,000 years for North America's placentals to begin diversifying, but over the following few million years they made an enormous rebound. During this time various lineages— such as the ancient hoofed mammals (condylarths), ancient primate-like creatures (plesiadiapiforms) and some bizarre, now extinct groups— rapidly diversified.[17]

The ancient hoofed mammals experienced the most spectacular evolutionary take-off. Just half a million years after the impact nine species were living in what is now Garfield County, the largest the size of a cat. Just one and a half million years later one of these animals, known as *Protungulatum*, had given rise to at least twenty new genera, the largest of which included species the size of a labrador dog. The group then proceeded to develop at the phenomenal rate of fifty new genera *every million years*! Four million years after the catastrophe it had given rise to a stunning diversity, including pony-sized species. Such rates of evolutionary change among mammals seem inconceivable today. They bespeak an empty world—the proverbial vacuum abhorred by nature—being progressively filled by that great bio-blob called Condylarthra, each new species filling a small niche until an entire ecosystem was created.[18]

In those days of vacuum-driven evolution the ancient hoofed mammals diversified into a variety of unlikely niches. One kind, the mesonychids, evolved into leopard-sized killers. Some still retained hoof-like toes despite their carnivory, while others developed pseudo-claws as well as sharp teeth. The largest and most specialised of the ancient hoofed carnivores evolved some fifteen million years after the catastrophe, and by then the world supported true carnivores, against which these ancient types must have successfully competed.

The forests inhabited by the condylarths were more dense than they ever had been during the age of dinosaurs. Many researchers have argued that this occurred because the enormous plant-eating dinosaurs had become extinct, allowing the forests to thicken up, much as some African woodlands change to dense forest when elephants are removed. Some palaeontologists think that these dinosaurs consumed the leaves of the ancient pines such as the araucarias, but those who study living ecosystems baulk at this, quite rightly pointing out that nothing living today can consume such tough and un-nutritious leaves. An organism can only subsist on very poor-quality fodder (such as leaves) if it has very low energy requirements (as koalas and sloths do), and/or is very large. Size is important, because large organisms can pass relatively greater volumes of poor quality food through their digestive systems. Low metabolic rates and relatively large size are the hallmarks of creatures that live on the edge—that is, those whose nutritional well-being is a delicate balance.

There is no doubt that the sauropods such as *Diplodocus* were very large, suggesting that they might have eaten araucarian leaves. Their teeth, however, look as if they were too weak to strip the tough leaves from the branches. Perhaps instead they cropped ferns from the forest floor and created gaps in the canopy by moving about. They might even have destroyed trees in order to create canopy gaps to encourage the growth of their favoured ferns. This would have been an invaluable strategy if they were territorial and stuck about long enough after knocking down a tree to benefit from the fern growth that followed.

Whatever the diet and habits of the dinosaurs, there is evidence that the increased density of forests after their demise encouraged some tree species to develop new strategies to propel their seedlings to the light before they starved to death in the gloom below. They began to produce large seeds packed with nutrients, which fed the seedlings until they could reach the sunlight. But this strategy had a drawback, for pirates were evolving in those forests. Foremost among these seed pirates were the condylarths, which systematically began to plunder this easy source of food.

The continued density of the forests may have prevented the early development of truly large condylarths. This is because seed-eaters are not particularly advantaged by size. With most of the foliage high overhead the leaf-eaters may have had to become arboreal, which of necessity places constraints on size, or remain small enough to survive on what little grew below.

Just two million years after the asteroid impact one group of condylarths—the phenacodonts—began its incredible rise in North America. In all, twenty-one species are known, eighteen of which are restricted to that continent, the remaining three having developed in Europe from North American ancestors. They were the first significant post-impact indigenous mammals to disperse throughout the continent and were among its most plentiful. All were herbivores or omnivores, varying from three to nearly 90 kilograms in weight. Some may have possessed short elephant-like trunks while other species had begun to develop physical differences between the sexes, including the weight and size of weapons such as canines.

For years it was thought that the phenacodonts gave rise to the perissodactyls, a group containing horses, rhinos, tapirs and many extinct species. Another group of condylarths, the arctocyonids, were believed to have given rise to an even more successful group, the artiodactyls, including sheep, pigs, cattle and their many relatives. As we will see later, much hangs on these hypotheses, for if they are correct North America can be thought of as having given rise, at an early stage, to virtually all of the larger herbivorous mammal lineages inhabiting the planet today. The continent could then be seen as a powerhouse of world evolution.[19]

By 57 million years ago the mammals had evolved into a distinct assemblage, differing from that of every other continent. The degree (if any) of immigration at this time is hotly disputed but, whatever the case, it was insufficient to prevent the development of a very distinctive North American fauna.

Among the most important tree-dwelling mammals of this fauna were the multituberculates. It is difficult to imagine these distant relatives of the platypus as they leapt from ancient pawpaw tree to cashew to avocado, seeking the nuts and fruit that comprised their diet in the tropical North American forests. Did they, one wonders, lay eggs as platypus do, and did they carry them in pouches on their daily rounds? The few marsupials present in these ancient forests are easier to imagine, for they were doubtless omnivores and small carnivores, just as most of their living South American relatives are today.

It was the placental mammals, however, that were growing in importance. If you had been able to look up into the trees of about 60 million years ago, you might have seen, alongside those leaping relatives of the platypus, the strange plesiadiforms, once thought to be distant relatives of our own species. Some may have resembled tarsiers, those curious, large-eyed insect-eating primates from south-east Asia which have inspired so many television gremlins. Others were more like the smaller lemurs of Madagascar. Below these tree-dwellers browsed the innumerable condylarths—some resembling dachshunds, others like pigs or tapirs. And between these, weird ground-dwelling placentals such as taeniodonts trundled about, none of which have left any descendants. The taeniodonts are of interest because they are only known from North America. They are rare animals in most fossil deposits, ranging in size from five to 110 kilograms; the smaller ones were omnivores but the larger ones probably grubbed for roots.[20] Their skeletons suggest that they looked like shrews grown to obscene size, though in real life they were doubtless more comely.

For all this rapidly evolving diversity, North America was still a post-apocalyptic world, inhabited by inept-looking smallish creatures lacking the breadth of form and size present during the age of the dinosaurs.

At the beginning of this first act we found a North America that had been devastated at the moment of its creation. Eight million years later, we see a continent living with that legacy. Its trees still lose their leaves in a tropical climate, its forests have altered in the absence of large herbivores, and its few animal survivors are rapidly filling the available ecological niche space. It is an utterly foreign place, as different from modern North America as is the island of Borneo today, yet in future millennia the continent would become even stranger.

Act 2

•

In Which America Becomes a Tropical Paradise

FIRST CONTACTS

Not all palaeontologists agree that, for the few million years after ground zero, contacts between the continents were severely limited. Some hold the view that North America, Asia and South America had extensive contacts immediately following the disaster, pointing to similarities between ancient kinds of mammals that existed on all three continents at this time in support of their argument. Such similarities, however, could also be seen as the result of convergent evolution.[1]

Whatever the case, as far as North America was concerned, near neighbours were never far away and the arrival of immigrants from Asia via Beringia or from Europe via Greenland seems at various times to have been a possibility. By about 57 million years ago—eight million years after the catastrophe—there is clear evidence that such an interchange had commenced, for strange creatures begin turning up in North America's fossil record. One of the earliest and most puzzling of these is a rabbit-sized animal called *Arctostylops*, which for decades constituted one of the most intractable mysteries in North American palaeontology.

The mystery began with the discovery, in the summer of 1913, of a tiny fossilised jaw—nineteen millimetres long—at a famous fossil locality known as Clark's Fork Basin, Wyoming. The specimen was found by one William Stein, a German immigrant employed to collect fossils for the American Museum of Natural History. When curator W. D. Matthew examined the specimen in New York he grew puzzled, then amazed, for Matthew believed the jaw had once belonged to a notoungulate, a primitive kind of placental mammal (whose name means 'southern ungulate') thought to be unique to South America.

The notoungulates and their relatives constitute a spectacular range of small to gigantic mammals. Some bore an astonishing resemblance to hippos, elephants and horses, even developing single-toed feet and high-crowned cheek-teeth long before the real horses developed such features. After surviving for 60 million years the last members perished just 13,000 years ago.[2]

Matthew pondered the possibility that the ancestors of these strange southern creatures had once existed in North America, but was confused to find that not a single other bone in Stein's Wyoming collection suggested the slightest connection with South America. As Matthew ruminated on the mystery he began to suspect that an awful mistake had been made, for Stein had been collecting in Argentina before he went to Clark's Fork Basin. Had the tiny *Arctostylops* jaw arrived in North America millions of years ago via a land bridge, Matthew wondered, or just a few months earlier in Stein's pocket?

The disturbed curator rushed a letter of inquiry to Stein, asking whether it was

> possible that you might have picked it [the jaw] up in Patagonia, wrapped it in cotton and tucked it in some out of the way pocket of your coat, and then missed it when you came to turn in your finds

at camp there. It might then stay in that pocket when you brought the coat back, and if you were wearing the same coat in the Big Horn [Clark's Fork] Basin, you might find it when you came to turn out your finds at night, and not remembering it, have supposed you had found it in the Wasatch Beds.[3]

To Matthew's relief, Stein promptly replied on 8 December 1913. His English was not perfect though his meaning was clear:

> Yours of the 2th came to hand today. I have been worried over the collection. I thought the quantity was too small, through my ignorance I did not know the quality of the fossils. Your kind letter was sure relieve to me, and I thank you for your congratulation. Yes we did work hard. We went in and over places wich was considerate impossible in summer time and drink water wich was thick with Wasatch clay!
>
> In regard to the specimen no. 79 it was find by me as stated on label and in record book by Mr Turner. I even remember the place. I find the two end pieces side by side but the mittle part was missing but after looking around a few minutes and utter some very strong word I was able to locate it a few feet below the others in a crack.[4]

It was not until 1969 that a second jaw was found, confirming that *Arctostylops* really had lived in North America. In 1924 similar fossils were found in Asia but until recently these discoveries simply deepened the mystery, as opinions about their significance were divided. Some researchers suggested that the fossil finds indicated that notoungulates arose not in South America, but in North America or Asia. Others defended the theory that the arctostylopids had migrated northward out of South America.

A few years ago a detailed study of these small fossils solved the puzzle.

It found that the North American and Asian creatures (including *Arctostylops*) are not notoungulates at all but a different family of mammals with superficially similar teeth. The arctostylopids were rather like hyraxes in size and shape, and possibly in lifestyle. Furthermore, discoveries have since shown that they were particularly abundant and diverse in northern Asia.[5]

Matthew would have been more astounded by the truth than by his suppositions, for Stein's tiny fossil had not come north by land bridge or steamer, but belonged to one of the first creatures to cross the Bering land bridge from Asia to America after the dawn of the age of mammals. The *Arctostylops* immigrants survived for just two million years, though they remain among the first of many immigrants to find a new home in continental North America.[6]

At about the time that the diminutive *Arctostylops* was making its way across Beringia, a much more formidable creature known as an uintathere (meaning 'Uinta beast', after the place of its discovery, the Uinta Mountains) was laying its first footfalls on the recovering continent. So odd are these primitive herbivores that they are placed in an order all of their own. They were the largest plant-eaters of their age, some being cumbersome beasts that weighed up to 4.5 tonnes. Despite their imposing size the origin of the uintatheres was for a long time a mystery as deep as that of *Arctostylops*. Their fossil remains have been found on three continents—Asia, North and South America—but their appearance in South America is brief and very early with only one kind, known as *Carodnia*, living there at the beginning of the age of mammals. There is even some doubt as to whether *Carodnia* really is an uintathere or some strange indigenous production—perhaps even a primitive relative of the notoungulates.

Whatever *Carodnia* was, there is evidence that the first true uintatheres evolved in Asia. By 58 million years ago very early specimens of a rather small size (twenty kilograms) had arrived in North America. The Asian side of the family remained modest in evolutionary terms, but their North American cousins embarked on a frenzy of diversification, sprouting horns and fearsome tusks and reaching titanic proportions. For 18 million years they would roam the forests of North America, the largest creatures of their age.

When I was a child reading my *How and Why Book of Prehistoric Mammals*, it was not the bloodied sabretooth that sent a tingle of fear down my spine, but a terrible-looking creature illustrated on the cover. It was not a predator (the artist drew it benignly munching on a mouthful of leaves) yet it was as big as a rhino with a bizarre head boasting three pairs of horns arranged from back of skull to nose-tip, and a pair of foot-long scimitar-shaped canines that would put the fangs of the most monstrous sabretooth to shame. The animal was an uintathere. What, I wondered, would it have been like to live in a world populated with such beasts?

Whenever I stayed up too late reading my *How and Why* books, I sometimes dreamed of meeting an uintathere in the thick dusty bush near my childhood home. It is impossible to move quietly in such a place, for each footstep crackles with the breaking of dry leaves and twigs. It's hot, and your skin and lungs are irritated by the dust. And then you see the great creature. You can never see very far in the dense tangle and you have stumbled across him from just a few feet away. As you turn to run, pursued by this nightmare of a beast, there's a vine to trip you up at every step. Spiny branches tear at your clothes before you wake drenched in sweat.

It turns out that I wasn't the only soul to be obsessed with these beasts, for their remains so impressed some of America's nineteenth-century palaeontologists that uintatheres were to change the course of the science on that continent. The first remains were discovered in the Washakie Basin of Wyoming in 1872, a time when there were still Indians, unbowed by the might of the American army, who freely roamed the ranges and plains. Joseph Leidy, founding father of American vertebrate palaeontology, had gone to Wyoming on his one and only expedition to the wild west and called it 'the most remarkable country I have ever seen'. Every member of the party carried a gun.

An awed and nervous Leidy spent days wandering through buttes 'that resemble great earthworks or huge railway embankments'. Some have 'the appearance of a vast assemblage of Egyptian pyramids'. On 17 July he entered even more formidable terrain, 'an utter desert, a vast succession of treeless plains and buttes, with scarcely any vegetation and no signs of animal life...An overwhelming silence reigned undisturbed even by the hum of an insect. Truly, I said, this is the wreck of another world which was once luxuriant with vegetation and teemed with animals'.[7]

The day ended in triumph for Leidy: before him lay the treasures of the desert, a huge pile of bones the likes of which the world had never before seen. He telegraphed a short note to the Academy of Natural Sciences in Philadelphia, breathlessly reporting upon the existence of 'the most extraordinary mammals yet discovered'. Most bones, he said, belonged to an enormous rhino-like creature, but there were great canine teeth in the rocks as well, some over a foot long, which Leidy believed were the 'most terrific instruments of slaughter', noting that 'their possessor was no doubt the scourge of Uinta'. Despite Leidy's conjecture that the teeth belonged to a great meat-eater, both canines and bones were from the same beast—the first uintathere to become known to the world of science.

Leidy was not the only palaeontologist in the field that year, for nearby was Edward Drinker Cope, and camped near Fort Bridger was Cope's nemesis, Othaniel C. Marsh. Cope was born in 1840 into a life of privilege and comfort. His father was a wealthy Quaker ship-owner, and although his mother died when he was just three (as did Marsh's), love was lavished upon him by an understanding stepmother and a close-knit family. Cope studied in Europe, then at the Quaker-run Haverford College in the US. After marrying his cousin he moved to Haddonfield, New Jersey, where as a man of private means he followed his passions in palaeontology and natural history at the Philadelphia Academy of Natural Sciences. Although he used the resources of the academy, almost all his life he worked as an independent scholar, producing close to 1400 scientific publications, many of them of considerable importance.

Cope was evidently an open and friendly individual who was nevertheless relentless in pursuit of his goals. His 1876 expedition in search of dinosaurs reveals how he tackled challenges. When he arrived at the frontier settlement of Helena in Montana, Cope found the town in an uproar, for Sitting Bull and his warriors had just destroyed the forces of General George Custer at Little Bighorn, and the white settlers lived in terror of scalping. Regardless of the risk, Cope set out with a scout, cook, two assistants, horses and a wagon, and pushed deep into Indian territory.

Cope believed that Sitting Bull and his braves had moved south and would not return to the badlands until late in the season. He was right about the majority of the Indian forces, but nevertheless he met small bands of braves, defusing these potentially dangerous encounters by popping his false teeth in and out, much to the amazement of the Indians. Still, the expedition must have been stressful, for Cope's assistant Sternberg wrote that 'when he went to bed the Professor would soon have an attack of nightmare. Every animal of which we had found traces during the day played with him at night, tossing him into the air, kicking him, trampling upon him. When I waked him, he would thank me cordially and lie down to another attack. Sometimes he would lose half the night in this exhausting slumber. But the next morning he would lead the party, and be the last to give up at night.'[8]

On the return journey Cope's cook and scout both deserted, fleeing in terror after spotting Sitting Bull's camp with its thousands of warriors. Cope and his two assistants were left to handle the massive fossils they had discovered. They reached the ferry-landing just in time to catch the steamboat *Josephine* on its last traverse of the Missouri for the year. Sternberg recalled that Cope arrived covered from head to foot in mud, his clothing hanging about him in wet, dirty rags. While a brilliant scientist and a resolute field worker, Cope was not money-minded. He poured his fortune into mining ventures and was left penniless when his investments failed four years before his death. He spent his last days in a rented studio, surrounded by his collections.

Othaniel Charles Marsh was born into a poor New York farming family in 1831. At first his prospects seemed bleak indeed, for he lacked direction in life. At a relatively late age he received funds from his uncle, the philanthropist George Peabody, whom he persuaded to support him while he studied at Yale. There, because he was so much older than the other students, Marsh was known as 'Daddy'. He further persuaded Uncle George to endow the Peabody Museum at Yale, where Marsh was given the post of professor of palaeontology.

Marsh's interests centred on the study of dinosaurs and ancient mammals. He was not a brilliant scholar (he once described the horns of the dinosaur triceratops as belonging to a kind of giant, extinct buffalo), but he was a thorough organiser and an astute manager. He surrounded himself with experienced and ingenious field workers, and to them is owed much of his success. It was his team, for example, which devised the method of throwing a plaster jacket over a fossil to transport it to the laboratory, a technique still used today.

Status meant a great deal to Marsh. In his later years he lived in a grand brownstone mansion in New Haven, where he entertained visitors in style. He was the first vertebrate palaeontologist to serve with the US Geological Survey, and for twelve years was president of the US National Academy of Sciences, an elevation that greatly irked the more talented Cope.[9]

To excavate and name such a grand creature as an uintathere is a matter of enormous scientific prestige. In mid-August 1872 these rivals hit paydirt simultaneously, and telegraphed back their own descriptions of the bizarre beasts, each imagining himself to be the discoverer. Leidy retired gracefully from the field, but Cope and Marsh fell into a festering rivalry that would break the health and reputations of both, and scatter enduring confusion over the name and nature of uintatheres. It would be more than a century before Leidy, his work ignored by both Cope and Marsh, would be recognised as the real discoverer, while the tangle of uintathere names created by the dispute remains problematic to this day.[10]

Marsh lacked some of the nobility and generosity of spirit Cope displayed, and this deficiency of character appears to be responsible for much of the unpleasant rivalry between the two. He encouraged his field workers to think ill of Cope, though after Marsh's field worker Arthur Lakes met Cope in Wyoming in 1879 he wrote, 'The *monstrum horrendum* Cope has been and gone and I must say that what I saw of him I liked very much. His manner is so affable and his conversation very agreeable.'[11]

It is hard to comprehend the intense ill-will that grew between Cope and Marsh in the years following their Wyoming fracas. In the end, it was as if their competitiveness towards each other had become more important than either the uintatheres or science. They produced lavishly illustrated manuscripts and published a plethora of redundant names in an effort to establish scientific priority and thus ownership of the group. Cope believed that the uintatheres were related to elephants but only managed to make the awesome beasts look ridiculous by placing elephant's ears on them. Marsh sniped that Cope was 'well known...for the number and magnitude of his blunders', and in 1886 published what he claimed to be the definitive monograph on the group. In 1889, stung by Marsh's coup and by a demand (which Marsh was partly behind) to turn his specimens over to the Smithsonian Institution, Cope returned fire. In a moment of intemperance he accused Marsh and his circle of being 'partners in incompetence, ignorance, and plagiarism'. The feuding did not cease until Cope's death in 1897, his reputation and self-esteem in tatters. Two years later Marsh died in a similar state, also financially ruined, leaving just $100 to his estate.[12]

But before all the sadness were the heady days of the 1870s, which saw palaeontology at its wildest and most magnificent, for these scientists were also required to be explorers. Charles Betts, a young student from Yale and one of Marsh's party left us an unforgettable picture of the expedition as it set off:

> We started from Fort McPherson escorted by a company of cavalry; for this was the country of the Sioux, and that warlike tribe was now in a state of unusual excitement. Across an unexplored desert of sandhills between the River Platte and the Loup Fork the celebrated Major North, with two Pawnee Indians, undertook to lead us. These guides rode about a mile in advance of the column. The major pointed out the least difficult paths; while the Indians, with movements characteristic of their wary race, crept up each high bluff, and from behind a bunch of grass peered over the top for signs of hostile savages. Next in line of march came the company of cavalry, commanded by lieutenants Reilly and Thomas; and with them rode the Yale party, mounted on Indian ponies, and armed with rifle, revolver, geological hammer, and bowie knife. Six army wagons, loaded with provisions, forage, tents, and ammunition, and accompanied by a small guard of soldiers, formed the rear.[13]

This amazing expedition was even joined briefly by that legend of the west, Buffalo Bill, who proclaimed Professor Marsh's geologising to be 'mighty tough yarns'. Each day the party pushed ever further into

Indian territory, in increasing danger of attack by the Sioux. Yet Betts wrote:

> we were not in the least alarmed when the Sioux really came into sight. Our composure was doubtless due to the fact that the warriors had been for some years dead, and were reposing on platforms of boughs, supported at the four corners by poles about eight feet high. On one of these tombs lay two bodies—a woman, decked in beads and bracelets, and a scalpless brave, with war paint still on the parchment cheeks, and holding in his crumbling hands a rusty shot-gun and a pack of cards. Beneath the platform lay the skeleton of the favourite pony, whose spirit had accompanied his master's to the happy hunting grounds. A feeling of awe crept over us as we built in through the historic castles of the dead, when the Professor brought us down to the stern realities of science with a remark: 'well boys, perhaps they died of small-pox, but we can't study the origin of the Indian race unless we have those skulls'.[14]

While Marsh was stealing skulls, horse-thieves were stealing his party's supplies. Marsh, unaware of this, pressed on to find enormous numbers of fossils, among which was the jawbone of an uintathere over a metre in length. Later the party crossed the Uinta Mountains: 'We stood on the brink of a vast basin, so desolate, wild, and broken, so lifeless and silent, that it seemed like the ruins of a world. A few solitary peaks rose to our level, and showed that ages ago the plain behind us had extended unbroken to where a line of silver showed the Green River.'[15]

Upon their return from the Uintas the party discovered that their supplies had been 'appropriated' by a band of suspicious-looking characters professing to be ranchmen, but whom the lieutenant accompanying them proclaimed to be notorious horse-thieves. Betts recalled that:

Professor Marsh went to the hut to claim our property. He was ushered into the presence of the party, each of whom was armed to the teeth, and looked ready to take his life for half a dollar. Endeavouring to control his embarrassment by speaking as to ordinary ranchmen, our illustrious chief remarked blandly, 'Well, where are your squaws?' 'Sir,' replied a dignified ruffian, 'this crowd is virtuous.'[16]

In these vivid accounts we see a west as wild as it ever was, a remarkable thing given that these expeditions took place less than 130 years ago.

While scientific opinion still wavers on the uintatheres' place of origin, evidence is growing that they, like the arctostylopids, evolved in northern Asia. Both groups may have crossed the Bering land bridge this early simply because they could tolerate the cool temperatures and long winter nights at such latitudes. And yet, despite their tolerance of cold, the ancient uintatheres fell victim to a decline in global temperature 40 million years ago. Before their extinction, however, they flourished spectacularly in North America. In some fossil deposits they constitute 20 per cent of the total number of species present and 50 per cent of the individuals recovered, suggesting enormous abundance and diversity. As a final testimonial to their success, just before their extinction, one North American species (a kind of *Uintatherium*) even managed to migrate to Asia and briefly flourish there.[17]

Asia was not the only source of immigrants at this time. Around 60–55 million years ago that distinctively South American group of lizards, the iguanids, appears in the north. Today, some forty-four species inhabit the continent north of Mexico, descended from several waves of invasion from the south. There is no need to invoke a land bridge to explain their presence in North America, however, for in recent years they have proved capable of long-distance sea travel. Their story, reported in *Nature* in 1998, is interesting for its own sake, as it shows just how effective oceanic dispersal can be for reptiles.

In 1995 a hurricane hit the island of Guadeloupe in the Caribbean's Lesser Antilles. Swollen rivers discharged vast rafts of flotsam into the sea. One tangle of waterlogged trees was used as a life-raft by no fewer than fifteen iguanas. For a month they floated on the sea until wind and currents washed them ashore on the island of Anguilla some 300 kilometres away. Anguilla was not previously inhabited by iguanas of this type, but once ashore they settled in and quickly began to breed. Perhaps a similar event brought the first iguanas to North America some 60 million years earlier, just as it brought them to isolated Fiji (over 10,000 kilometres east of South America) a million or so years ago. Given this early dispersal and the ease with which reptiles can make such journeys, it is surprising that there was no further dispersal of reptiles and amphibians from South America until about 35 million years ago, at which time tree frogs (*Hyla*) invaded North America from the south.[18]

Between 57 and 55 million years ago a motley collection of mammal immigrants made the journey from Asia to North America, including the ancestors of rats and their kin and some very odd-looking beasts. Imagine an animal—say a hedgehog or a shrew—that has adapted to eating vegetation and grown to the size of a cow and the shape of a hippo. There you have *Coryphodon*. Even 55 million years ago, when it first arrived in North America, it was a relic of a bygone age. Despite its superannuated appearance *Coryphodon* did well in its new home. Its bones, heaped into piles, are so common in some deposits that they are termed 'Cory dumps' by the palaeontologists who search rocks of suitable age in America's west.[19]

The cause of their migration was a gradual warming trend that continued for 14 million years after the asteroid impact. There were slight ups and downs in global temperature, which corresponded to various immigrations and periods of isolation, but overall the warming made the Bering land bridge more hospitable for many creatures. Nothing terribly dramatic occurred until about 51 million years ago, when the warming trend peaked in a heatwave, the likes of which had not been experienced for over 60 million years. For a brief period (perhaps just a million years or so) in the early Eocene, temperatures reached extraordinary highs. The mean temperature of the North Sea hovered around 28 degrees Celsius. Even the deep oceanic waters of the South Atlantic checked in at about 12 degrees Celsius while at the poles the sea surface temperature was about 14 degrees. There seem to have been no cold places on Earth at all, unless the summits of the highest mountains provided icy refuges, evidence of which has not been preserved in the fossil record.[20]

The cause of this unprecedented though short-lived event may have lain in the arcane movements of the continental plates across the surface of the Earth. This caused increased volcanic activity in east Greenland, and outpourings of lava on the sea floor as the continents rifted apart. The result was a huge increase in the amount of carbon dioxide in the atmosphere. The sea itself began to circulate in a different way at this time, for differing saltiness in the various oceans began to drive currents that brought warm, salty water to the poles, only to see it cool and descend into the depths to be reheated at the equator.[21]

The planet had become a true greenhouse world, whose warmth would open the continent's polar portals and allow massive immigration between North America and Eurasia. Such conditions have not been repeated in the 50 million years since, but industrial emissions may ensure that in coming centuries the world will again enter such a greenhouse.

THE BRIDGE OVER GREENLAND

I tend to think of the heatwave of 51 million years ago as the time when North America forgot itself, for never again would the continent be so dramatically out of step with its normal state. For more than a thousand millennia ancient lemurs sported in the treetops of deciduous forests growing within the Arctic Circle, while goannas and Australian turtles lurked below. Further south, where there was no polar night to bring on seasonal change, North America was covered with evergreen tropical jungle. Imagine the badlands of North Dakota clothed in the jungles of Panama. It's easier if you look in the rocks, for there you see the fronds of ancient Dakota palms and leaves of extinct relatives of the avocado, the cashew and the soursop.[1]

These ancient North American rainforests abounded with pangolin-like anteaters and thus probably with termites. From timber to termites to anteaters and the eaters of anteaters, this was a strange ecosystem. At its apex were not carnivorous mammals but enormous flightless birds with beaks like poleaxes. At 1.8 metres tall, *Diatryma gigantea* was a rail grown monstrous, capable of killing and swallowing whole most of the North American mammals of its day, for only three genera exceeded the gigantic avian predator in size, the primitive coryphodons among them.[2]

Not surprising given the tropical conditions, reptiles and amphibians flourished throughout the continent. The sirens, an obscure family of eel-salamanders with forelimbs but no hindlimbs, leave their first fossil record. So too do the mole salamanders, a group that includes the tiger salamander—the largest land-dwelling salamander in the world. Both salamander families are unique to North America and may well have longer, albeit undocumented, histories on the continent.

The land tortoises (Testudinidae), a family that includes the familiar gopher tortoise, appear for the first time as immigrants from Asia. They arrive in North America, one palaeontologist wryly notes, two full geological stages before the hares, anticipating Aesop's fable by about 50 million years.[3]

Today one must travel to remote reaches of Arnhem Land in Australia's Northern Territory, or the Fly River of New Guinea, to see a pig-nosed turtle, but 50 million years ago they could have been seen basking along many of the rivers of North America. Just how this bizarre living fossil (imagine a large, soft-shelled turtle with flippers like a leatherback and a flexible snout like a pig) came to survive in this obscure corner of the world is one of biology's more trivial yet intriguing mysteries.

Those distinctive carnivorous lizards, the goannas or monitors, also stalked the Eocene jungles of North America. The group had become extinct in North America with the asteroid impact 15 million years earlier, but now they returned, probably via the Bering land bridge. Relatives of Australia's frill-necked lizards (agamids) were also present in North America at this time. Although both lizard families are extremely diverse in Australia today (75 per cent of all living goanna species, for example, are found in Australia) both have their origins in the Old World, and both became extinct in North America soon after the heatwave passed.

The tenure of goannas in North America is puzzlingly brief, for they are such ubiquitous and successful lizards elsewhere, able to fill ecological niches occupied by weasel- to coyote-sized carnivores. Whatever the reason, their cameo appearance 51 million years ago was their only North American role during the entire age of mammals.

Skinks and iguanids were doubtless also present but, while both are familiar American reptile groups today, they remain all but invisible in the fossil record laid down during the heatwave. The North American knob-scaled lizards (xenosaurids) and night lizards (xantusiids), however, were more common and were found further north than the few remnant species (which are largely Central American) appear today. Crocodiles were also abundant—no less than seven genera of which lived alongside five kinds of alligator. No such crocodilian diversity exists anywhere today.

One remarkable effect of the heatwave was that it opened a veritable highway to Europe. Given the current configuration of the continents, it seems absurd that North America and Europe could come into intimate contact while leaving Asia relatively isolated. But then the relative positions of continents were very different. The North Pole lay directly over northern Beringia, the land bridge joining Asia and North America. Despite the warmth this stymied migration between these two continents. North America and Europe, however, were joined by overland connections at less northerly latitudes.

One problem we have envisaging the world of 51 million years ago concerns the nature of Europe itself. The very *idea* of Europe is a fossil, both of our colonial and of a deeper geological history. A glance at any atlas reveals that Europe is not a separate continent at all, and it has not been since the Early Oligocene period 32 million years ago, for that is when it became welded to Asia, and its fauna was overwhelmed by that of its larger neighbour. During the earliest Eocene (55.5 to 52 million years ago), however, Europe was a distinct continent bounded to the south by the warm and shallow Tethys Sea and to the east by the Obik Sea, which divided it from Asia. Even then Europe was the smallest of the continents and its geological independence the most fragile. It is more accurate to think of Europe at this time as an island archipelago with polar connections to North America rather than as a separate, coherent continent. Yet it did have its own distinctive flora and fauna—hedgehogs that looked like huge bristly rats with bony shields on their foreheads, primitive insectivores and early hoofed mammals, to name just a few.

We know about the bristly fur of the hedgehogs because of a fossil find made near Messel in eastern Germany. There the preservation is so exquisite that frogs have been found still bearing traces of their original colours, while fossil jewel beetles still sparkle as brightly as the day they were entombed 50 million years ago. Even the bacteria inhabiting the nasal cavities of various creatures have been preserved. Recent excavations of the site (once earmarked to become a garbage-dump) have revealed a wide variety of birds, mammals, reptiles, amphibians and even insects, fish and plants.

The Messel site preserves many species of American origin, such as primitive rodents (themselves recent arrivals from Asia), diatrymas, marsupials and hoofed mammals. They had arrived courtesy of the heatwave, which made several land bridges habitable. Some doubtless crossed the more hospitable southerly route known as the Thule land bridge. It connected North America—via Greenland and the Faroes—with the British section of the European archipelago. It opened only briefly some time between 50 and 55 million years ago, and was then severed by continental rifting. The more northerly DeGeer route, connecting North America (again via Greenland) with Scandinavia, remained open until about 46 million years ago. This route was not discovered by palaeontologists until 1971. It is named after the great fracture in the Earth's crust called the DeGeer Line, along which the North American and European plates slid past each other as they finally separated some 45 million years ago.[4]

It is hard to imagine all those exotic creatures traipsing across Greenland and the Faroes, but these barren wastes must have then been richly forested and productive, for a plethora of reptiles and mammals once lived there. Intriguing evidence suggests that the ancestors of some now familiar birds might also have used these paths, for biochemical studies reveal that starlings and mockingbirds are related. Mockingbirds, those delightful North American creatures that behave like wind-up toys, first evolved in the early Eocene, most likely from ancestral starlings that flew over the DeGeer route, this time crossing from Asia via Europe to North America.[5]

What was the result of this first joust between the inhabitants of North America and Europe? Did the primitive plesiadiforms and the archaic coryphodons of North America prevail, or did Europe's superannuated rat-like hedgehogs presage the events of 1492? On this the fossil record is clear—it was a hands-down victory for the invading North American forces. Terrifying diatrymas stalked the DeGeer route into Europe along with primitive fox-terrier-sized horses, ancient ancestors of tapirs, rhinos and primitive relatives of the camels, yet very few Europeans went the way of the mockingbirds-to-be. As this diverse horde took up the new lands on their eastern frontier the old Europeans died out. The ancient hedgehogs were to be no more. Ultimately the migrant Americans evolved and changed in their new island home. One camel relative (known as a caenothere), for instance, ended up looking (and possibly behaving) like a hare, a group then absent from Europe, while the rodents evolved into curious and large forms.[6]

By the end of the interchange, at least half of the sixty-one

land-mammal genera in Europe had close relatives and common ancestors in North America, principally as a result of Europe acquiring a
new fauna. Such close faunal similarity between these two continents
has never been seen before or since, despite the best efforts of acclima-
tisation societies, farmers, and other lovers of introduced species.[7]

There is one slight twist to this story of North American dominance:
the warm conditions that opened the DeGeer Route to the North American invasion of Europe also allowed some migration across the Beringian
land bridge. Many researchers are now asking whether or not this was
an early manifestation of the domino effect, with immigration starting
from Asia then flowing into North America, then Europe. In other words,
the North American victory may have been won by newly arrived Asian
immigrants. This view is not supported by the fossil record, however,
for among the successful invaders of Europe were unique American
creatures such as the phenacodonts, that group of early ungulates that
prospered after the asteroid impact.

It was during the period of faunal interchange in the early Eocene
(about 55–50 million years ago) that many modern kinds of placental
mammals first appear in the fossil records of Asia, North America and
Europe. Among the most important are the artiodactyls (cattle, sheep,
hippos, etc.), perissodactyls (horses, rhinos and tapirs) and early dog-
like carnivores. The great riddle is, which of these three continents gave
rise to the ancestors of such familiar beasts as cattle, horses and dogs?

For decades, students of palaeontology were taught that the
thoroughly North American phenacodonts had given rise to the perissodactyls, but new fossil finds in Asia have driven certainty from this
dictum by suggesting instead that this landmass was the homeland of
the group. The origin of the artiodactyls is also now clouded with doubt.
Opinions have ping-ponged between Asia and North America about
the point of origin as one new fossil find follows another. Recently,
however, the discovery of the skeleton of the primitive hoofed mammal

Chriacus, a rabbit-sized creature that lived 60 million years ago in New Mexico, has once again given North America the advantage.[8]

While on the subject of North American origins, it is worth jumping ahead 10 million years to an important subgroup of artiodactyls, the Ruminantia (those that 'cheweth the cud', such as cows, sheep, deer). They were long assumed to be Eurasian in origin, yet new, strong evidence for this important group makes a compelling argument that they first appeared 41 million years ago in North America.

These seemingly obscure scientific debates about animal origins are of great importance to us, for they strike at the heart of our question— what sort of place is North America? Has it always been an 'ecological superpower', exporting its superior fauna and influence? Or has it been a net recipient of creatures and plants evolved elsewhere? In other words, is the present American dominance of global interests out of step with its deep history?

Had I been writing this book only a few years ago, I would have portrayed North America as the principal fountainhead of many successful mammal groups, a land of emigrants that successfully colonised the world. This was once the prevailing view, which came not from a full understanding of the global fossil record, but from the precocity of palaeontology in North America. Because this science has flourished on that continent, its disciples have discovered many ancient remains widely believed to represent the first appearance of many important mammal lineages. Now a more complete Eurasian record is known, and it indicates something very different. Christopher Beard of the Carnegie Museum in Pittsburgh has developed a scenario based on new evidence that he calls the 'East of Eden' hypothesis. It's an idea first championed by Dr Matthew (he who sent Mr Stein into the field) many decades earlier. This title, with its biblical allusion, neatly encapsulates the idea that many groups of mammals first thought to have originated in North America did in fact evolve near the traditional location of mythical Eden, Asia, before finding a home in North America. Matthew's view was even more reminiscent of the biblical story, for he believed that humanity itself arose in this Asian Eden.[9]

One of biology's more iron-clad rules seems to be that the inhabitants of larger lands are likely to be more successful immigrants than those of smaller ones. One possible explanation for this is the idea that because larger landmasses support more species than do smaller ones, competition between species is fiercer and evolution progresses at a faster rate. A widespread pattern of speciation known as centripetal evolution seems to bear out this idea. Centripetal evolution constantly generates new species at the centre of a group's range, leaving relictual species around the margin, often on islands.

Rapid evolution can of course occur at the margin, such as when a species first reaches an offshore island. Then one sometimes sees truly spectacular rates of evolution as birds become flightless or creatures such as elephants become pygmy versions of their former selves. This evolution is quite different from that occurring on the continents, for on islands species adapt quickly to environmental constraint while on the great continents competition from other organisms seems to be the driving force of evolutionary change.

These considerations lead us to ask whether North America is a 'centre' or 'periphery' in terms of mammalian evolution. Because the continent has changed so much through time, we must ask the question repeatedly. Before we do this, however, we must first establish North America's position, in terms of its size, in the pantheon of continents.

Today, Eurasia is by far the largest continent. Its 54 million square kilometres—more than a third of all land on Earth—dwarf its rivals. Next in size is Africa at over 30 million square kilometres, constituting about 20 per cent of all land. North America (24 million square kilometres and 16 per cent) is third in place, followed by South America (nearly 18 million square kilometres and 12 per cent), Antarctica (13 million square kilometres and 9 per cent), and finally Australia (7.6 million square kilometres and just over 5 per cent). I do not think that size is everything in determining how the inhabitants of continents fare during periods of faunal exchange, but I do concede it is, as Beard, Matthew and the contemporary scientist Jared Diamond postulate, a powerful explainer of success for larger, warm-blooded creatures.

Have the ancient fluctuations in the size of continents had any discernible influence on North America's ability to operate as a biological cornucopia, exporting its fauna worldwide 50 million years ago? Until 33 million years ago Europe was an island archipelago, its 10 million square kilometres being quite separate from Asia but intermittently connected to North America and in time forming a single biological and geological unit with it. Including Greenland, this 'Euramerica' (North America + Greenland + Europe) formed a continent well over 36 million square kilometres in extent. Until about 40 million years ago, India (more than three million square kilometres) was a detached fragment of Gondwana, adrift in the Indian Ocean. Thus, 50 million years ago, North America and Asia were much more nearly equal in size, since Asia (minus India and Europe) constituted roughly 40 million square kilometres in extent, while 'Euramerica' would have covered at least 36 million square kilometres. It was a warm land, and much of its area was forest-covered. This makes it seem much more probable that at least some of the now dominant groups of modern mammals which first appeared about 50 million years ago did indeed arise in North America.

The fossils needed to test this idea, however, will probably be long coming; looking back on this part of the age of mammals and trying to make sense of it is like tracking a single strand of pasta in a bowl of spaghetti. There are many fossils from just a few locations, and most look rather similar, for they were all evolving with great speed from just a few basic types. It is, I think, too early to answer the big questions about where most of the now-dominant groups of mammals came from. A little later in the record, however, we begin to see clearer patterns and it is to this period that we must turn for answers about the nature of this strange continent.

Here closes the second act in the great play of North American pre-history. In it we have seen America establish relations, via land bridges, with the rest of the world. A period of unprecedented global warming saw the continent's northern portals thrown wide open; its fauna undertook a wholesale and highly successful invasion of its smaller neighbour, Europe. Indeed, for most of the second act Europe functioned as an appendage of North America. The warming also allowed strange faces to be seen in North America—Australian pig-nosed turtles and goannas among them. Not until the twentieth century would such a wholesale invasion of temperate and tropical adapted species again stream into the continent.

Act 3

•

In Which America Becomes a Land of Immigrants

Six

A FATAL CONFIGURATION

Charleston, South Carolina in January can be a reminder to visitors that North America is a great thermal trumpet. Everywhere in the coastal forests are signs of the tropics—groves of palmetto, broadleaved trees, even ferns. Yet the air temperature during the day can be as low as a chilly 4 degrees Celsius, dropping below zero at night. The cause is frigid air, freight-expressed from the Arctic. Its presence in a tropical paradise tells us that North America is a unique place and that one of its most distinctive features is its climate.

North America's climate is strongly determined by its shape—a great, inverted wedge with a 6500-kilometre-wide base deep in the sub-Arctic. To the south the wedge narrows until it is reduced to a peninsula just sixty kilometres wide, terminating eight degrees north of the equator in a narrow isthmus boldly abutting an expansive South America. On its eastern side the wedge is reinforced by the Appalachians, while in the west the Rocky Mountains perform the same function.

No other continent has this configuration. The closest match is Asia, which is also wider in the north than in the south but, as Jared Diamond has noted, it differs in one profound way from North America: its mountain ranges run predominantly east–west. With the exception of the otherwise forgettable Ouachita Mountains of western Arkansas, North America is cursed with mountain ranges that run north–south.[1]

Land cools and heats more rapidly than water; thus land temperatures vary more than those at sea. In winter, air that has become super-chilled over America's vast northern expanses surges southwards, funnelled by the north–south ranges towards the tropics. In summer a huge pool of air warmed over the Gulf of Mexico surges in the opposite direction, bringing the tropics to the far north. This 'climatic trumpet' created the 'express fall' and 'express spring' experienced over much of the United States, and its chilly blasts afflict sun-seeking January visitors to Charleston.

The continent's shape and highland orientation thus have a pro-digious impact, which is compounded by the fact that North America has barely shifted in latitude—just ten degrees south—over the past 80 million years. Because of its latitudinal immobility the changing world climate has played across the land's surface like wind upon a field of wheat, blasting its flora and fauna north as the earth has warmed, and south as it has cooled, ultimately exiling forever many ancient tropical species as their zone of comfort disappeared over the southern horizon.

When Earth is warm (in greenhouse mode)—as it was around 50 million years ago—North America is a verdant and productive land. Almost all of its 24 million square kilometres, from Ellesmere Island in the north to Panama in the south, is covered in luxuriant vege-tation. But, as the Earth cools, North America's capacity to amplify change rapidly drives it to a break point, beyond which it falls into the frigid grip of the poles. It can then be said to be in icehouse mode, a mode that characterises the present. The last shift to extreme icehouse conditions occurred around 18,000 years ago, then returned to the present (still ice age) conditions just 10,000 years ago.

Today about eight million square kilometres (a third of the continent) is barren Arctic tundra or boreal forest. Drop the deep-sea temperatures just another two degrees Celsius, and North America becomes dominated by a field of ice some 18.5 million square kilometres (75 per cent of its present land mass) in extent. Understanding this great determinant of life provides a useful tool in comprehending the forces at work across the continent.

Temperature ranges in North America can vary incredibly over a brief period. The town of Spearfish in South Dakota holds the world record, going from −18.9 degrees Celsius to 3.3 degrees above in two minutes.[2] Spearfish is nestled between the Black Hills and the rolling plains which seem to stretch endlessly to the north. It is vulnerable to storms generated in the Arctic or in the nearby brooding ranges, but was a tranquil place on the morning I gazed out at it from the windows of the Valley Cafe, which was abuzz with the news that 'Rooster Rost' would be performing that evening. A slow June drizzle softened the hard edges of the town that grew upon the banks of a shallow stream where the Sioux came to spear fish. The Black Hills were their last refuge until gold was discovered in the 1870s. Ten thousand miners then descended on the area and Spearfish, which today boasts 6966 residents, was born. It still has the feel of a frontier town, its main street boasting a trade store (Sharp's Trading Co.—gold bought and sold), the Spearfish Bootery, and the Pawn Doctor Inc. Even the ordinariness of death that so characterised the American frontier seems not to have quite deserted this place, for a rolling plaque advertiser above the door of the cafe carries gaudy advertisements for cars, furniture and real estate, as well as one for the Fidler Funeral Chapel—Randy Rimington, Director.

As I left Spearfish I was alarmed to find a police car, lights flashing, parked across the road ahead. For a moment I thought that something more exciting than the weather had happened in the little town. Then I saw a line of cars turning down a lane towards the cemetery, with another police car bringing up the rear. Randy Rimington was busy.

The continent's climate means that conflict between the north and south not only characterises American history, but North American prehistory as well. Across the land, turbulent air flowing from the chilly north encounters the breezes of the hot south. As the two fight it out over the plains, tornadoes are spawned. Ninety per cent of the world's tornadoes occur in North America—most originating between the Rockies and the Mississippi River.[3]

The North American climatic trumpet plays two tunes. One is seasonal, being responsible for winter's chilly blasts and summer's heat. The second, a longer note, is played out over geological time, shifting the continent from greenhouse to icehouse modes. A small shift in global climate—in effect a breath of cool air elsewhere—is magnified until its effect is doubled or tripled, calling forth glaciers and fields of ice. There is one North American biome that is so distinctive and speaks so eloquently of this phenomenon that it begs for first consideration—the great deciduous forests that reach their apogee in New England. Here, we must begin as the forest did, with nuts.

I never understood nuts until I wandered through the forests of New England. Before that I got them in a packet: walnuts, pecans, hazels. What really intrigued me as I munched my way through the delicious kernels was the idea that a tree would go to all the trouble of producing a nourishment-filled nut—and then not defend it properly. The shells of North American nuts are relatively thin. Not at all like those of Australian macadamias or South American brazils, two of the few commercial nuts that are not from the northern hemisphere; their shells are so tough you need a mallet to get into them.

Then I watched the squirrels. 'Rats with fancy tails' a friend calls them—but to me they seem wondrous. Some live their whole lives under the gaze of humanity and they do such interesting things. It was fall when I first saw them and they were busy carrying nuts everywhere. Holes in fences, tree trunks and stones were stuffed with them. Lawns had tiny excavations clawed into them to receive yet more nuts. And then I realised. These trees *want* the squirrels to eat the nuts. They work on the principle that squirrels aren't perfect; that in all this activity one of them will forget where it stored some of its nuts, get run over by a car or be made into stew before it can retrieve its stash. Thus, while many nuts are eaten, some are carried into the perfect environment to nurture a young tree.

The trees are competing with each other to have the squirrels carry off their nuts; during a short interval in the fall the ground is littered with nuts that they have not deigned to squirrel away. Thousands, nay millions, of generations of squirrels have been busy shaping those nuts by carrying off, carefully hiding and sometimes forgetting those that appeal to them the most. I think that nuts that are easy for squirrels to collect, to open and that taste good will be given the royal treatment and hidden away first. They prefer nuts and acorns that are thin-shelled and easy to open; some are even shaped so that they are conveniently carried in a squirrel's mouth, and it is these that, if they are not eaten, stand the best chance of growing into trees.

In the southern hemisphere there are no squirrels except recent immigrants, and they do not need to cache nuts to survive their mild winters. Mammals in the southern hemisphere usually eat the nuts where they find them, thereby destroying the reproductive potential of the tree. Brazils and macadamias armour their nuts accordingly.

Squirrels, I should point out, appear to be a truly North American phenomenon. They evolved there some 30 million years ago at a time when North America had already become a cool, seasonal land, and ever since they have been adapting to local conditions, evolving new forms to suit the ever-changing opportunities the continent offered. They have on occasion crossed over into Eurasia, but these early immigrants did not flourish. It was only about two million years ago that modern tree squirrels really took hold there. Curiously, attempts to transplant squirrels to places like Australia have almost all failed.[4]

Native Americans have much to thank their squirrels for; they prepared the ground for the development of unique American Indian cultures. From California to the Mississippi Valley, from the Pacific Northwest to Florida and right through to the deciduous forests of New England, lived Indian tribes that were dependent upon nuts as a source of food. These were nuts which squirrels had been selecting for size, hardness, taste and nutritional value for millions of years. Like the squirrels, these Indians gathered and stored the nuts, which were often present in such numbers that they encouraged the development of complex societies, sometimes consisting of permanent towns with thousands of inhabitants. Only in North America do we find complex societies existing away from the coast in the absence of agriculture. They are a unique expression of the synergy between the climatic trumpet, deciduous trees, squirrels and some remarkable human cultures.

The leaves of North America's deciduous forests tell an even more intriguing story. As I write, I am sitting atop a tower named after the Indian-hating Puritan, Increase Mather. The tower rises sixty metres above the New England landscape and below me are the streets of Cambridge, and the Charles River snaking towards its barrier. To the west rise gentle hills covered in bright green foliage which, even now in early September, shows the slightest bronzing indicating the change to come. The trees are going to sleep. Saying that trees lose their leaves is misleading. Loss suggests an inadvertent process: 'Oops, I've lost my leaves!' It is a far more deliberate strategy than that, for trees *throw away* their leaves, first draining them of nutrients, leaving brightly coloured waste products that are the glories of fall.

Even in profligate North America little is lost to the tree when it sheds its leaves, for a leaf loosed into the New England fall air has a 99 per cent chance of landing within twenty to thirty metres of its source. As its leaves rot in spring, it's quite likely that the tree will be able to recoup whatever investment in nutrients it put into making the leaf, just at a time when it needs it most.[5]

Forests dominated by deciduous trees thrive wherever the average temperature of the coldest month of the year hovers about the freezing point or less, but where that of the warmest month exceeds a balmy 20 degrees Celsius; in other words, wherever there is a truly tropical summer followed by a chilling, Arctic winter. As we have seen, this is just the kind of seasonal pattern that characterises North America. Deciduous trees also dominate in Europe but the more stable climate there gives them much less latitude for diverse, flamboyant displays and restricts their distribution.

Where summers do not reach such balmy heights, evergreen conifers tend to take over, and where winters are not so chill the evergreen broadleaf forests dominate. In North America's south-east, however, deciduous forests thrive where evergreen broadleaf forests should grow, forcing the evergreens into the subordinate role of understorey shrubs or small trees. This anomaly results from blasts of frigid air blown from the climatic trumpet. They wither the leaves of the evergreens and turn the water in their vascular systems to ice. As the plants thaw, embolisms of air form, giving them the vegetable equivalent of a heart attack. Their deciduous neighbours are untroubled by such things, for their vascular systems are protected by their architecture. They sleep on, ready to assume dominance across vast tracts of North America, just as they have done each spring for much of the past 65 million years. These trees are the vegetable kings of climatic extremes; but relax the extremes and the evergreens outcompete them. Such climatic extremes are almost entirely absent in the southern hemisphere, as are deciduous trees. [6]

Many deciduous trees have the same, star-like (palmate) leaf pattern.

Maples (*Acer*), planes (*Platanus*), sweet gums (*Liquidambar*), basswoods (*Tilia*) and tulip trees (*Liriodendron*) are just a few separate plant families to have independently evolved this leaf shape, as too have some oaks. Such leaf shapes, I hasten to point out, are very rare in evergreen forests. Why should the deciduous trees have evolved such a shape over and over again?

Star-shaped leaves, it turns out, are marvels of economy. The radial pattern of supports with a thin web of green tissue spread over them is probably the most economical way to construct a leaf. But it is a leaf not meant to last. For six months or so it does its job, then it is discarded like a hamburger wrapper or a paper coffee cup.

If we are to understand how this very distinctive American forest took shape we must leave the present and delve again into the distant past, to a time between 40 and 32 million years ago when North America first took on its modern aspect. I encountered this crucial moment in Earth history in one of the most enchanting landscapes North America has to offer—the wide, rolling plains of Wyoming.

SEVEN

LA GRANDE COUPURE

In the fall of 1983 I was an impoverished student attending my first international conference, the forty-second annual meeting of the Society of Vertebrate Paleontology in Laramie, Wyoming. I don't think I've ever had so much fun. I drank at bars with cowboys, gaping in astonishment at the racks by the bar entrance on which they dutifully hung their guns. I went on field trips to important fossil localities out on the range and saw for the first time American antelope, coyote and prairie dogs. It was as I scanned the grasslands, squint-eyed, trying to pretend that all those cows were so many buffalo that I first saw the legendary palaeontologist Morris Skinner. Dressed in a stetson and a shoestring tie with a clasp of blue stone set in silver, the octogenarian Skinner cut a dashing figure. He was pointed out to me by an awestruck young devotee who whispered that in the days when Morris 'rounded up' his fossil horses, it was still necessary to pack six-shooters to protect oneself against outlaws and the like. I never found out the truth of such statements, but they suited the romance of a man who has done so much to reveal the exceptional history of this continent to the world.

The Wyoming rocks in which Morris Skinner and his loyal assistants searched for fossils encompass an extraordinary expanse of time. After the bridge over Greenland existed, they continue upward in a sequence that can be clearly seen in the bluffs and spires of the badlands. The eight million years they cover, from about 40 to 32 million years ago, was a critical time in the evolution of North America's modern fauna, and without the lifetime of effort that Morris Skinner put into collecting and painstakingly preparing fossils, we would know far less about it.

Skinner's work is all the more remarkable because he was employed by the Frick Laboratory, a research centre described to me by a colleague as something of a 'mysterious operation'. Childs Frick, a millionaire many times over, was obsessed with fossils. He established his laboratory in the 1920s, and while his father Henry, partner of Andrew Carnegie of America's steel empire, and sister were accumulating some of the world's finest art on the east side of Manhattan's Central Park, Childs was busy compiling a collection of fossils that rivalled any held in any public museum in the world. Frick died in 1965 and in 1968 his collections were donated to the American Museum of Natural History, just across the park from his sibling's art collection. Even today his fossils comprise about 80 per cent of the museum's fossil mammal collection, and occupy seven floors of its ten-storey Frick Wing.

Unfortunately, Frick seems to have been more concerned with the quantity of fossils collected than which stratigraphic level they came from, and his collectors seemed to have set their targets—Skinner among them—by the boxload. Yet Skinner soon realised that the fossils were worthless without precise stratigraphic information, for sometimes just thirty centimetres up or down a rock sequence made a million years' worth of difference. All of his fossils are meticulously documented and thus became a legacy that would change our notions about an entire continent.

During the 20 million years of stratigraphic time over which Skinner roamed, two periods of very dramatic cooling occurred, the first at 50 million and the second at about 38 million years ago. Sediments reveal a global drop in the temperature in the deep sea of about 4–5 degrees Celsius in each episode. By comparison, the deep-sea temperature dropped by only two degrees Celsius around 25,000 years ago as glaciers stole over the north of the continent. As usual, North America magnified each cooling; on the Gulf Coast the mean annual temperature dropped by about nine degrees Celsius, while a six-degree Celsius drop occurred in southern California.[1]

As one might imagine, such massive drops in temperature had dramatic effects. Hard winter frosts were felt in the high north for the first time and the lush, tropical rainforests that covered much of North America began to open out as a result of drying as well as cooling. The new seasonality gave deciduous trees an advantage and they began to replace the evergreens wherever the seasonal difference was most extreme. In the far north, such as the Gulf of Alaska, it was the bitter winter that gave the deciduous species the advantage, but elsewhere winter drought may have been the decisive factor.[2]

As the cold began to take hold about 50 million years ago, rainforest started to vanish from the south-east of what is now the United States and was replaced by deciduous species of the thorn tree family (the legumes), along with oaks, walnuts and laurels. They formed a community rather like the tropical deciduous forests that now lose their leaves in response to the long winter dry season in parts of Mexico. The importance of these changes cannot be overestimated for they indicate that dramatic seasonality was finally establishing itself over the continent.[3]

In the American west the change in climate was accompanied by changes in the land itself that would in turn accentuate those wrought by climate. After 25 million years the Rocky Mountains had ceased to push skyward. From now on, they would slowly sink amid the burden of their own erosional debris, and by 20 million years ago they had been reduced to insignificant flattened summits poking up from a high plain of sediment.

The pressures within the North American crust that had built the Rocky Mountains now moved on and were forcing one piece of the Earth's crust below another. When this process occurs, the crustal rock being forced down is melted in the Earth's mantle and returns to the surface through the vents of volcanoes. Volcanic activity was thus renewed on a vast scale throughout the western cordillera, and a series of volcanic uplands began to develop. Volcanoes of the explosive St Helens type were particularly active at this time. Some boggle the mind with their immensity, including one in the Needles Range of Nevada and Utah that resulted in the ejection of over 2100 cubic kilometres of pulverised, superheated rock into the atmosphere. Ashfalls of this size must have affected the global climate for decades, perhaps producing a mini nuclear winter of the sort that began our story. In time eastern Oregon, Nevada, western Utah (then a flattish land that was repeatedly blanketed in volcanic ash) and other parts of the continent would receive rejuvenated soils as the ash decomposed.[4]

During this period the Rocky Mountains region supported a mixed and changing vegetation. In one part of the Wind River Basin of Wyoming, which dates to the early part of the period (over 50 million years ago), there were palms as well as poplars, herbs and vines, many of which were deciduous. In other parts of the basin the vegetation included conifers, horsetails and over forty species of flowering plants, 60 per cent of which are thought to have been deciduous. Again, the development of a long dry season rather than a hard winter seems to have been responsible for this deciduousness. In the Green River Basin, famous for its fossil fish, which was laid down a few million years later, remains have been found of a seasonally dry deciduous forest similar to those growing today at 1000 metres elevation in the mountains of south-western Mexico.[5]

Despite these changes, the American west was to act as a refuge for several spectacular plant species, foremost of which is the Sierra redwood (*Sequoiadendron giganteum*), a stately survivor from pre-impact times. The higher peaks of the Sierras provide the last redoubt for these largest and arguably most extraordinary trees on Earth. In 1983 I undertook a pilgrimage to view these giants, leaving Los Angeles for a long drive north, first viewing the great coast redwoods in the Coast Ranges north of San Francisco before crossing the Sacramento–San Joaquin Valley to see the eastern side of the Sierras. I drove ever higher through the dry mountains until the temperature cooled and mist filled the air. Somehow I missed the road sign to one of the largest living organisms on the planet, which was a fortunate mistake, for I found myself instead in a car park surrounded by other majestic trees—the largest being the Douglas fir (*Pseudotsuga menziesii*). Today the Douglas fir and sequoia co-occur at just two places: in the Placer County Grove east of Foresthill and the Tuolumne Grove near Yosemite. The co-occurrence is a rare and magnificent relict, for the association was far more common in earlier ages.[6]

A path led off through this avenue, a sign assuring me that the sequoias lay in that direction. As I walked through the enchanted grove I quite forgot about the trees that were my destination, for the life of the forest was abundant and consumed all of my attention. Squirrels, deer and myriad species of birds caught and held my gaze. All seemed framed between dark, columnar trunks of pine and fir, rich humus below and hovering mist above. Then out of the mist a titanic reddish-pink column appeared. At first I could not comprehend the size of this structure, nor exactly what it was. It absolutely dwarfed the Douglas firs that had captured my imagination as giants. It was a sight forever etched on my memory, as was my struggle to grasp the full significance and grandeur of what was before me—one of the largest living things on Earth.

I approached the tree with reverence, as quietly as I could, then beside it out of the mist appeared the trunk of another, then another giant. The grove consisted of perhaps a dozen trees. As I neared them I heard human voices breaking the stillness of the forest, yet I could see no people. It was not until I reached the bole of the monarch that I found the source of the noise. The road I passed came up to the very base of the first tree and vehicle after vehicle was slowing then speeding off. Many of the tourists did not deign to alight; their obeisance consisted of shoving a camera through an open car window, pointing it and pushing the button before driving away.

While redwoods are now indubitable symbols of the American west, 45 million years ago they also grew in Europe. The stump of one specimen, over nine metres across, was recently discovered in a German coal mine. Soon after this tree grew the bridge over Greenland was severed by vast movements in the Earth's crust, isolating the European redwoods from their American brethren.

A few million years after that great German sequoia fell, the Beringian land bridge would again open a way from North America to Asia. Although Asia was home to many advanced species it still supported a large number of archaic creatures. One such living fossil that I would not have stood in the way of was a primitive Asian carnivore known as *Andrewsarchus*. Only the skull of this terrifying animal is known, but it is about ninety centimetres long and sixty centimetres wide and has a wolf- or bear-like appearance. If its proportions were similar to those of a grizzly, *Andrewsarchus* would have been four metres long and two metres high at the shoulder, making it the largest mammal land carnivore of all time, but in the absence of more bones such an estimate is pure speculation.[7]

Despite its archaic inhabitants, Asia was a dominating influence on

North America between 45 and 35 million years ago, for the continents were intermittently connected and most migration was one-way—into America. Many palaeontologists argue that North America acquired its first true rhinoceroses, peccaries, beavers, pocket gophers and many other rodent types, as well as larger now-extinct animals, during this period. The rhinos were to prove spectacularly successful, radiating into a variety of niches. Until the elephant-like mastodons arrived 17 million years ago they were the largest mammals in North America. There were at least four major immigrations, each one introducing new rhino species that generally ousted the old, until the last American rhinos succumbed to climate change and competition from other herbivores about 3.5 million years ago.

Not all of the dominant mammal groups at this time were immigrants from Asia, for some native North Americans were doing very well indeed. One such group is the oreodonts. During their entire period of existence (46.7 to 5.2 million years ago) they were restricted to North America. These sheep-sized creatures were distant relatives of the camel and unexceptional-looking. In some fossil deposits, however, their bones comprise a quarter of all those found, suggesting an astounding dominance. Indeed, their remains are so common in some strata that they have given rise to a new and lucrative industry: you can buy a beautifully preserved oreodont skull at many rock shops in America for a few hundred dollars.

The lifestyles of the oreodonts have been a mystery for some time. Some possessed eyes on the top of their heads like hippos, which certain researchers have taken to indicate an aquatic life. Oreodont remains, though, are most common in windblown sediments, indicating dry conditions. New and still contentious studies—focusing on well-preserved remains of animals that were presumably buried where they lived—suggest that some oreodonts may have been burrowers. Some skeletons even have the remains of foetuses, usually two, three or four, preserved in the mother's belly. Such large animals tend to have so many young only if they live a precarious life, prompting one researcher to suggest that oreodonts used those eyes atop their heads to peek over the rims of their burrows before emerging. But what kind of danger were they keeping an eye out for?[8]

The caution of the oreodonts may well have been prompted by a horrible new arrival from Asia. Pigs are loud, smelly, and have a propensity to grow warts and bristles in unseemly places. Their feeding habits are none too savoury either, for although they are herbivores by ancestry they have a liking for meat. Usually they scavenge, but when they get the opportunity they will kill. The immigrants that may have worried the oreodonts were a now-extinct group of pig-like creatures known as entelodonts, which seemed to have all of the pig's most unpleasant attributes in abundance.

My introduction to entelodonts was through the fossil collections of the Museum for Comparative Zoology at Harvard, where cabinets are crowded with the remains of diverse, long-extinct animals. Atop one (it's too big to fit inside) lay the skull of one of the largest crocodiles ever found and on top of another the skull of an early whale. The friend who was showing me the collection opened one drawer with the casual question, 'Can you guess what kind of beast this is from?' Before me lay a narrow skull the length of my forearm, from the sides of which sprang prominent, arched cheekbones. It looked like it was once part of some carnivorous, marine creature, and I was about to venture this guess when my companion turned it over. There, in two rows, were the unmistakable pig-like teeth, complete with a set of fine tusks. It was an entelodont. Drawer after drawer contained their remains: jawbones sporting great bony warts the length of my thumb, fearsome tusks dwarfing those of any razorback, and skull fragments from giant-sized porcine animals. My friend then opened the bottom drawer of the cabinet, where the heaviest specimens reside, and I beheld a reddish, nondescript lump of bone the size of my head. At first I had no idea what it was, for it was badly broken and rounded, but soon I saw the base of a huge bony wart and the socket for a tusk that must have been the size of a walrus's. Here lay the sad, broken remains of the mother of all entelodonts. In life, the creature must have been the size of a small bull, its skull alone a metre long.

Scott Foss of Northern Illinois University has recently completed a detailed study of the unfriendly entelodonts. These beasts had an enormous gape that could fit around very large objects and their skulls are often covered in healed wounds, including crushed snouts and shattered facial bones. These wounds were made by teeth suspiciously similar in size and shape to awesome entelodont premolars. Those massive jaws, it seems, were designed to bite down hard on other warty entelodont heads.

Foss suggests that the entelodonts were scavengers of the carcasses of large creatures like rhinos, but I don't think they would have passed up scoffing an oreodont as it came to the entrance of its burrow to greet the day. If they did, it is certain that the hyaenodons, archaic carnivores resembling dogs or bears, also recent immigrants from Asia, wouldn't have missed the opportunity. Had we lived 30 million years ago, 'as fearful as an oreodont' might have been a household maxim.[9]

A very curious thing was happening in the slowly drying, ever more seasonal forests of North America 40 million years ago. Among the myriad life forms scrambling about on the forest floor was an animal about the size of a small fox. *Hesperocyon* was a common creature that was fated to give rise to the entire dog family. For over 30 million years the dogs and foxes were to remain restricted to North America, their subsequent spread making them, as we shall soon see, one of very few successful North American fauna exports.[10]

At the same time that *Hesperocyon* slunk through the Eocene undergrowth of North America, a group of cat-like carnivores called nimravids was rapidly diversifying. But these creatures from Asia were to die out without leaving issue, while the real cats evolved in Eurasia. Was it just chance that determined that North America should be the home of dogs and Eurasia the cradle of cats? I think not, but that story must await its proper place in this continental history to be comprehensible.[11]

Other soon-to-be important groups of herbivores were differentiating in this changing North America, foremost among which were the horses. Their story is told in the fossil record with such exuberance that it is impossible to ignore, which is perhaps why Morris Skinner took such an interest in them. You get a sense of their importance when you visit the palaeontology collections at the American Museum of Natural History in New York. Upon entering the elevator of the Frick Wing one is reminded of a surreal department store. Beside the button for each floor is a list of names: level 5, for example, is devoted to oreodonts and the related omomerycids. Horses, however, have an entire floor to themselves.

By 45 to 50 million years ago, horses had established themselves as a significant element in the North American fauna. Although they had begun to diversify, the various kinds remained dog-sized rainforest dwellers and eaters of fruit, seeds and herbs. Then the continuous forest canopy of the greenhouse world began breaking up, providing new opportunities in more open habitats as the icehouse took over. Over time the plains would come to the horses, but it was a tortuous process, with the very changes that would ensure the horse's ultimate success also threatening it in the meantime. By 40 million years ago, just a single species of horse, of the genus *Epihippus*, survived, and it makes only a rare appearance in the fossil record. It looked as if the group might become extinct before a larger leaf-eating kind adapted to the drier, more seasonal conditions. Slowly, however, they did adapt and soon horses once again abounded in North America. One wonders how different human history would have been if *Epihippus* had not made its knife-edge escape from extinction.[12]

This period also saw the advent of the earliest known camel, a creature called *Poebrodon*, known only by a handful of teeth from 46-million-year-old rocks from Wyoming. This humble animal would leave a spectacular array of descendants—forty-four distinct genera in total, almost all of them unique to North America.[13]

The life-changing cooling episodes of 50 and 38 million years ago, however, just hinted at what was to come, for 33 million years ago a cooling so dramatic occurred that it closed the lengthy Eocene epoch. The Palaeogene extended for 41 million years, from 65 to 24 million years ago, and is divided into three geological epochs: the Palaeocene (ancient), Eocene (dawn) and Oligocene (slight epoch, because it is so brief). Palaeogene, means 'ancient period' because the mammals that lived then seem so alien to the modern world. It is followed by the Neogene—the time of the new, which is itself divided into the Miocene (the 'less new' epoch), the Pliocene ('more new') and Pleistocene (the 'most new'). The prodigious cooling event that opened the Oligocene was so abrupt that Hans Stehlin, the Swiss palaeontologist who first recognised signs of it in rocks, called it *la grande coupure*—the great cut.

The great cut may have occurred over just a few tens of thousands of years. Various causes for this have been cited, including massive volcanic eruptions—even errant asteroids—but recent research points to the movement of continents as the cause, the fault lying largely with Australia. Beginning 60 million years ago, Australia began to unzip itself from Antarctica. By 45 million years ago overland connections had been severed but a long submarine rise stretching south of Tasmania still joined the two continents. By 33 million years ago even this final, tenuous connection had been sundered and a deep ocean had opened, allowing bottom water to pass between the continents for the first time.

Today, 233 million cubic metres of water pass between Antarctica and Australia every second. This is one thousand times as much water as flows down the Amazon in an equivalent time; and the water is frigid, for it constitutes the Antarctic Circumpolar Current. Scientists think it was the establishment of this current that turned the planet into an icehouse world, and it did so by keeping cold water and air circulating around Antarctica, thereby generating the cold bottom waters of the world's oceans and fostering the establishment of the Antarctic ice cap. Other factors, such as the opening of the Norwegian Sea and the Drake Passage (between South America and Antarctica) may have played some part in the great cut, but the establishment of the Antarctic Circumpolar Current was the switch that turned on the modern world—a switch thrown by a wandering Australia.[14]

The *grande coupure* is also the moment when a decisive shift in the world 'balance of power' occurs. India had joined Asia after a Himalaya-raising collision 40 million years ago. Then Europe, by now isolated from North America, finally joined the accreting Asian landmass. With these additions, the new Eurasia dwarfed every other continent including North America. But worse, the ability of North America to amplify climate change meant that cooling had robbed much of its north of productivity. After this there would be but one 'ecological superpower': Asia.

At the time of the great cut, mean global temperature dropped by an extraordinary five to six degrees to just 5 degrees Celsius. North America, however, experienced a stupendous decline of eight to twelve degrees Celsius in mean annual temperature. This cooling had a profound effect both on land and sea. Along the Gulf Coast of the United States, some 97 per cent of marine snails such as conches and whelks, and 89 per cent of clam species, became extinct. Imagine walking along a warm, sandy beach and stopping to pick up a few of the hundreds of species of brilliantly coloured shells washed onto the shore. Then picture yourself returning to that beach a few thousand years later. The temperature is more than ten degrees Celsius cooler, about the average annual difference between New York and Miami, and you find that more than nine out of ten kinds of shells have gone, leaving just the cold-tolerant ones. This was the reality of North America 32 million years ago.

The forests were similarly affected. Studies of leaves reveal that before the great cut parts of Alaska experienced a mean annual temperature of about 15 degrees Celsius, with an annual range of ten degrees Celsius or less. After the cut, the mean annual temperature dropped to just 7 degrees Celsius. But the leaves reveal a far more important piece of information, for they tell us about seasonal variability as well. What the Alaskan sequences show is that, after the cut, the annual range of variability was about ten degrees Celsius on *either side* of the annual mean! Seasonal variability had increased fivefold. The tropical summers and hard cold winters typical of so much of the continent had finally arrived in North America.[15]

As a result of these climatic changes, the mixed deciduous-evergreen forests of Alaska gave way to mixed northern hardwood forests composed of alders, beeches and hickories. Even that ancient survivor of polar climes from the age of the dinosaurs, the dawn redwood, made a comeback. It must have been hiding out in some refuge during the millions of years that greenhouse conditions persisted, perhaps right near the pole, awaiting a change of fortune. The return of this deciduous tree, which is tolerant of a mean annual temperature of just 7 degrees Celsius, speaks eloquently of how devastating the change was.

Throughout North America, wherever plant fossils of this age are found, similar changes are documented. The deciduous plants were once again on the move, dominating in region after region. This time seasonal variation rather than drought forced them to throw away their leaves. In the Pacific Northwest, the kinds of plants now growing under the mighty redwoods of California and Oregon were becoming prevalent, replacing the subtropical broadleaf forests that preceded them. In Mississippi, Alabama, and as far north as New Jersey it was the oaks, adapted to cooler and drier conditions, that were rising to dominance.

The cooling had a profound effect on those ancient survivors, the reptiles and amphibians. There was a dramatic decline in the once-abundant numbers of freshwater tortoises, crocodiles and alligators. In their place the land tortoises prospered, for they are capable of withstanding both dryness and cold.

Many of the archaic mammals characteristic of earlier ages had been in decline since the Earth began to cool 50 million years ago. Very few survived the chill of the great cut. The last of the multituberculates, those squirrel-like relatives of the platypus, died out a couple of million years before the big freeze, but many other groups persisted right up to the crisis. Most notable among these were the brontotheres, great rhinoceros-like creatures that had dominated the large mammal niche up to that time. Their low-crowned teeth tell the story. They were adapted to eating browse, and high-quality browse was hard to find in this dry, cold world. Along with the brontotheres went the last archaic rodents, some insectivores, distant relatives of the camel and certain primitive moles.[16]

Europe was being transformed by the great cut, too—though for somewhat different reasons. Following the severing of the bridge over Greenland 45 million years ago, Europe had become an isolated island archipelago. But then the rising of the European Alps and a fall in sea level that accompanied the great cut dried the Obik Sea, opening Europe to Asia. Most of the archaic creatures that had inhabited Europe since the North American migrations of ten to 15 million years earlier were exterminated as new Asian species invaded. The army of Genghis Khan was not the first it seems, nor even the most devastating Asian force to enter Europe, for at the time of the great cut 32 million years ago a more mixed Mongol horde (including ancient bears, raccoons, weasels, mongeese, bear-dogs, nimravids, rhinos, peccaries, entelodonts, beavers, squirrels, hamsters, mice and rabbits) arrived, transforming Europe and making it their own. They swamped the old Europeans (most originally from North America), driving many to extinction and restricting the distribution of others. Compared with *la grande coupure* experienced in Europe, the North American fauna got off lightly.

A curtain of ice, or at least very cold conditions, descended on Beringia 32 million years ago, bringing about a cessation of faunal interchange with Asia. North America was to go it alone for a while, an isolated continent in the grip of great climate change. The only groups to flourish during this harsh time were canids, horses, oreodonts and camels—all distinctively North American families that had evolved on the continent—and the rodents. Mammal diversity in North America reached its nadir, for not since the dawning of the age of mammals were there so few species filling such a limited number of ecological niches on the continent.[17]

At some stage during this long and increasingly chilly history, a most curious dispersal occurred: the ancestors of those adaptable, cold-loving attendants of death, the ravens, arrived in North America. They, along with the crows and jays, had doubtless crossed via the frozen land bridge, but their journey had begun at a far more distant point. Ravens are, it seems, one of Australia's very few successful animal exports, and their extraordinary world-girdling journey began on that southern continent at least 35 million years ago. It led them first to Eurasia, then via Beringia into North America and finally some two to five million years ago into South America, giving them a global distribution. Given Australia's role in creating the climatic catastrophe in the first place, there's something appropriate about these flying emissaries of death originating in the Great South Land.[18]

Birds, incidentally, have left a rather poor fossil record, so it is difficult to trace their origins and dispersal using fossils. Instead, scientists have resorted to molecular studies. In the 1960s researchers learned that, if boiled briefly, DNA would 'melt', separating into its two separate strands. One of these strands can then be labelled with a radioactive isotope and mixed with an unlabelled strand from another species. If you incubate this mixture at a temperature of 60 degrees Celsius for 120 hours the strands combine, forming hybrid DNA. The mixture can then be reheated to determine at what temperature the two strands separate. The greater the similarities between the two strands, the more firmly they stick together. This gives a rather elegant overview of how similar the DNA of various species is, and thus how closely they are related.[19]

Such DNA studies have dismissed many erroneous hypotheses. Earlier taxonomists (myself included) had previously assumed that turkeys constituted an endemic family of North American birds, for they are such distinctive-looking creatures and so clearly North American in origin. The biochemists have proved us all wrong, however, by showing that turkeys are nothing more than glorified pheasants. While unique to North America, they have evolved from European pheasant stock in relatively recent times.

Biochemical studies reveal that very few, if any, bird families have originated in North America. The only current candidates are the new world quail (family Odontophoridae), which includes such familiar species as the bobwhite. All are rather poor fliers with a limited ability to disperse. The North American pedigree of even this group, however, is doubted by some biochemists, who comment that they 'must be the descendants of an early divergence in South America while that continent was isolated from North America'.[20]

In addition to the ravens, those warty creatures the true toads along with the back-fanged snakes, arrived in North America during the great cut. The toads (family Bufonidae) along with tree frogs (family Hylidae) somehow drifted across the sea from South America. The back-fanged snakes, which evolved in Europe some 32 million years ago, had reached North America by 30 million years ago, probably via Asia. Today they are the dominant snakes of North America, including such familiar creatures as the racer, garter snake, corn snake and king snake. In all, ninety-two of the 115 snake species now found in the United States and Canada belong to this family (Colubridae).[21]

EIGHT

A GOLDEN AGE

All felicitous tales tell of redemption, and the story of North America is no exception. Following the gloom of the great cut a new dawn was experienced; the sun shone warmer and America's plains were once again to fill with life. The Neogene—the time of the new—was on hand and the continent was about to enter a golden age.

The Miocene period, which immediately follows the time of the great cut, encompasses a long and complex 20 million years of Earth history— from 24 to five million years ago. In the Miocene the planet warmed, but strangely rainfall did not increase as it had in previous warm phases. Dryness seems to have prevented the broadleaf forests from returning to cover the continent and instead large tracts of North America became a wooded savanna.

North American fossil deposits of this age regularly yield the remains of fifteen or more species of mammals weighing more than five kilograms, a diversity never seen before or since in North America and in our time existing only in the richest grasslands of east and south

Africa. By ten to 15 million years ago, when large mammal diversity reached its peak, a standing crop of between seventy and eighty genera of large mammals (weighing more than five kilograms) inhabited the continent. Compared to today's twenty-two (which includes such small animals as foxes and marmots) this was a grand total indeed.[1]

The Miocene can be divided into three phases. During the first, which lasted about five million years, there was rapid diversification. This climaxed during the second phase, while the third saw a slow decline, again lasting about five million years. The age was a sort of window of opportunity that opened then closed in response to altered global climate patterns, themselves due in part to vast tectonic changes that included the re-emergence of the Rocky Mountains from under its mound of debris and the uplift of the Andes and the Himalayas. The result was a warming and drying climatic trend, providing just the conditions that grasses need to thrive. The closure of this 'window of diversity' resulted from a continuance of the drying trend coupled with climatic cooling— once the optimum point was passed the world (and North America in particular) became less productive, being too cold and dry to support a cornucopia of mammal giants. By 4.5 million years ago, the species diversity of mammals had dropped to near its ice-age level of 20,000 years ago.[2]

Fifteen million years ago, America's Serengeti boasted locally evolved ecological equivalents of most of Africa's 'big five'—the elephant, buffalo, black and white rhino and lion—as well as smaller game. The elephant niche was filled 17 million years ago when the first members of the elephant group (proboscideans)—the mastodons and later the gomphotheres—reached North America via Beringia from Asia. The living elephants are known in ecological parlance as 'keystone species' because they alter the environment in ways that allow smaller creatures, such as antelope, to thrive. The arrival of elephant-like creatures in North America just before the height of the golden age may well have changed the environment in ways that were favourable to mammal diversity, thus enhancing the existing trend. They may, for example, have opened holes in the forest canopy, allowing meadows to exist or providing opportunities for pioneer plant species, which would in turn favour greater animal diversity.[3]

The African black rhino, a browser on low-quality fodder, found its North American counterpart in a rhinoceros named *Aphelops*. A more unlikely relative was another North American rhino known as *Teleoceras*, which resembled the hippopotamus. Despite their different ancestries these species bore a close resemblance to each other, for they were the same size, had similar limb proportions and even similar life strategies. The lifestyle of *Teleoceras* has been documented by Dr Michael Voorhies, who excavated wonderfully preserved skeletons at a place called Poison Ivy Pocket near the town of Orchard, Nebraska. Ten million years ago a massive volcanic eruption occurred somewhere in the mountainous west of the continent. The huge avalanche of pulverised rock it released drifted out over the Great Plains, blanketing hundreds of square kilometres in a thick layer of abrasive ash. At what is now Poison Ivy Pocket, the ash formed a layer three metres thick as it filled a pond. The rhinos died in their waterhole, the ash blanket preserving their remains in a pristine state. While excavating the hyoid bones in the throats of these ancient rhinos, Voorhies found fossilised seeds of the ancient grasses they fed on. His discovery proved that these great, short-legged rhinos lived in water but ate grasses that grew on the plains beyond, just as hippos do today. Because an entire herd was killed Voorhies was able to examine the population structure and determine the rate of infant mortality and found that it resembled those of hippos. *Teleoceras*, as hippos do, lost appallingly large numbers of young (up to 45 per cent) to predators, whereas land-based rhinos lose few, usually around 8 per cent.[4]

Other fossils preserved at Poison Ivy Pocket give us a glimpse of the grandeur that was North America in the Miocene, for the remains of birds, three-toed horses and elephant-like gomphotheres accompany those of *Teleoceras*. Because they died where they lived the preservation is exquisite; the birds even have impressions of feathers surrounding their skeletons. The emerging picture at Poison Ivy Pocket is of a place similar to what one would see around a waterhole on Africa's Masai Mara.

Despite the importance of elephant-like species and rhinos in the American Miocene, it is the native Americans—the horses, camels, oreodonts and other extinct groups like the 'slingshot-horned' protoceratids and the dromomerycids (rather silly-looking deer-like animals, some with forward-curved horns)—that comprised the bulk of the herbivore species. The camels reached their finest flower at this time. One, the appropriately named *Aepycamelus giraffinus*, bore an uncanny resemblance to the long-necked creature and may even have exceeded the largest modern giraffes in height, being at least three and a half metres tall at its shoulder. Others camels resembled gazelles, while yet others were more akin to llamas and dromedaries.[5]

During the middle to late Miocene (20 to ten million years ago) the number of North American horse genera doubled every five million years. At the height of their diversity no less than twelve different kinds of horses, including both browsers and grazers, roamed this American Serengeti. All were three-toed and apparently one species of the genus, *Neohipparion*, even undertook mass migrations as zebra still do today. Imagine looking out over the present-day African savanna and seeing twelve different kinds of zebra. It's not such a far-flung fancy, for all zebras are descendants of North American ancestors, and the North American fossil record tells us that such diversity once existed. In places like Florida at least nine species co-existed, while in the Love Bone Bed of northern Florida horse remains comprise nearly 60 per cent of all ungulate fossils, suggesting an exceptional abundance.[6]

The Miocene also saw horses make their first major breakouts from

North America. After more than 20 million years of evolution in isolation they finally dispersed, at least three times, to Eurasia and Africa. Their principal centre of diversification nevertheless remained North America, for the immigrants failed to flourish and were replaced by later waves of horse immigration. At or just after the end of the Miocene camels also emigrated; certainly they had arrived in Eurasia by four million years ago.

Another endemic group appeared about 19 million years ago when the ancestors of the pronghorns (which today form the only family of large mammals still unique to North America) arrived. They are true products of the Miocene, arising from Eurasian ancestors near the beginning of that period and suffering a great decline by its end. Their precise relationship with other ungulates is unclear, although both bovids (cattle, sheep, antelopes) and deer are possible close cousins.

The first pronghorns belonged to a group which the great palaeontological philanthropist Childs Frick christened the 'pronglets' (merycodontine pronghorns) because of their small size and graceful proportions. The dog- to sheep-sized pronglets so beloved by Frick diversified and became widespread during the second phase of the Miocene, while the antilocaprine pronghorns (to which the living species belongs) became important during the third phase. Some extinct antilocaprines were very odd, exhibiting pairs of corkscrew-like horns on their head, while others had three pairs of horns set at different angles.

Today there is just one survivor of this golden age of grazers—the incomparably swift and enduring American pronghorn (*Antilocapra americana*). John Byers, the latest researcher to study the species, writes that the biology of the pronghorn is a 'historical document that illuminates the fossil-written text of the North American savanna fauna'. In the pronghorns' biology Byers reads a story of the lost and golden age of the Miocene.[7]

Byers makes the observation that 'pronghorns are ridiculously too fast for any modern predator and are faster and apparently have greater stamina than Thompson's gazelles'. Their extraordinary speed—up to 100 kilometres per hour—and their considerable stamina bespeak a creature that evolved in clear and present danger. That danger, Byers thinks, came in the form of the continent's extinct meat-eating megafauna, which included cheetahs, running hyenas and short-faced bears. Byers thinks that the pronghorn's reproductive behaviour was also shaped on the huge, megafauna-filled prairie that was once North America, for the females hide their fawns assiduously, even eating their faeces and drinking their urine so that predators can smell no trace of them. Given the land's current dearth of predators such extreme caution seems unnecessary, but in a landscape filled with megafauna it may have been essential.

Why the America savanna fauna of 15 million years ago should so closely resemble that of present-day Africa is an intriguing question. Some researchers have sought explanations in the idea of co-evolution. They say that each species on Earth is shaped by interactions with the other species in its environment and that on the savanna there are only limited choices available as the various species compete. Large browsers, they believe, have to be shaped like giraffes in order to reach food in the treetops, while grazers are best off being fast and migratory like horses, semiaquatic like hippos or enormous and well-armoured like white rhinos. Other researchers dispute this argument, explaining that any similarities may be due to coincidence, yet whatever force is at work must be powerful, for it is able to create the same community 'shape' over and over again in response to similar environmental conditions, using entirely different animal stock.[8]

Despite the enormous diversity of its mammals, the first phase of the Miocene is notable for one absence; it is the only time in 37 million years of evolutionary history that North America has lacked cat-like creatures. The 'cat-gap', as palaeontologists call it, extended for at least six million years from around 23 to 17 million years ago. Before this, a group of cat-like carnivores (the nimravids, which were very like cats though only distantly related to them) thrived. Some were partly arboreal, ranging in size from house cat to leopard, while others were enormous, with sabre-shaped canines, like the true sabretooths of a later age. Why did these cat-like creatures die out in North America (while surviving in Eurasia) with no replacement by the true cats? Their fate may be owed to the same factors that created the diversity of herbivorous mammals, for most cats need forest or cover from which to hunt. In an increasingly open America the nimravids may have found themselves without an ecological perch to hunt from, particularly if competition with dogs prevented them from colonising the savannas. The dogs, incidentally, responded to the extinction of the nimravids with the evolution of several cat-like forms which became extinct when true cats arrived from Asia 17 million years ago. Even today North America is home to a somewhat cat-like dog in the grey fox, which is an adept climber. The earliest true cat immigrants were about the size of a lynx, but soon evolved into a variety of species, some the size of a lion.[9]

Some six million years after the first true cats arrived in North America, one last line of nimravids, the genus *Barbourofelis*, entered the continent from Asia. Commonly known as vampire cats they were among the strangest creatures ever to tread the face of the Earth. The first thing you would have noticed about *Barbourofelis* was its flattened, sabre-like twenty-centimetre-long upper canines. On opening its jaws you would have seen six pointed lower incisors, each as large as a coyote canine, projecting sharply upward. The only cheek tooth was a solitary slicing blade, leaving its owner unable to chew its food. This unusual dentition indicates that *Barbourofelis* was a hypercarnivore, able to eat nothing but flesh. It may even have specialised on the softest tissues such as innards and blood. If its teeth inspire terror, the rest of its body only excites mirth, for its limbs are pathetically short and its body long. It must have resembled a large-headed, sabre-bearing bulldog. It is hard to imagine it making a kill: perhaps it lunged at the necks or bellies of short-limbed rhinos from ambush.

Members of the bear, weasel and raccoon families, as well as gigantic 'bear-dogs', shared the plains with the cats and dogs of this ancient continent. As we have seen, dogs (family Canidae) are a true production of North America, a group that evolved there and has spent most of its evolutionary history restricted to that continent. Most larger dogs have evolved into pursuit carnivores, superbly adapted to running down game on the open plains. Cats, however, are stealthy attackers, preferring to hunt from a concealed location. The cheetah is an exception to this rule and the only genus of living cats that may be North American in origin, possibly being related to pumas (you can see resemblances in the small head and slender body). Cheetahs moved out of North America and into Asia and Africa a couple of million years ago, becoming extinct in their homeland just 13,200 years ago. What is a cheetah but a cat trying to be a dog? The constraints of life in North America may well have fostered such a marvellous transformation as this.[10]

Other Asian intruders that arrived at this time include more elephant-like mastodons and gomphotheres, several kinds of rhinos, and bizarre creatures resembling clawed horses called chalicotheres. In a three-million-year period, between 18.8 and 15.8 million years ago, no less than sixteen genera of Eurasian mammals established themselves in North America. Carnivores comprised the most diverse group of immigrants—including bears and the returning cats.

At the other end of the scale and a little later in time, new kinds of mice (the cricetids) arrived, spawning a great radiation of these ubiquitous creatures. By the end of the Miocene, one rodent species that was to have a profound influence on the European colonisation of North America had arrived. The first modern beavers had swum the streams and ponds of Beringia and began dam-building in their new continent.

Other kinds of beavers had long been resident in North America, which has an ancient tradition as home to aquatic rodents, though none have survived to the present.[11]

In examining the fate of just one of these immigrant groups we can gain important insights into the nature and effect of such large-scale immigration. The story concerns the three bears. Not papa, mama and baby, but a less lovable threesome: the primitive bears (amphicynoids), running bears (hemicyonines) and modern bear (ursines). These are the three basic kinds of bear that have evolved over the past 40 million years. All three first evolved in Eurasia, and all have invaded North America at various times. For a while each kind was wildly successful there, but as new bears arrived from Eurasia the older types became extinct. The replacement was not instant or abrupt, but a slow dwindling of the old species as the new diversified. This pattern is seen in group after group, for rarely does the arrival of new immigrants spell the instant destruction of old endemics. It is also important to note that, despite the evolution of some truly astonishing bears in North America (including the largest mammalian carnivore ever to stalk the Earth), no North American bear has ever successfully invaded Eurasia.

The running bears are the ecological equivalents of the larger dogs. They invaded North America at least twice during the Miocene, and competed with members of the dog family, but not until the very end of the period did those true North Americans, the dogs, invade Eurasia. The world itself would have to change before native Americans such as dogs would find their place in it.[12]

Throughout this golden age North America was geologically remodelling itself. Activity was, as always, greatest in the west with the Great Basin and Pacific coast being the major foci. Nowhere did this geologic activity create such an eerie landscape as in the basin and range country of Nevada, Utah and the south-western states. When nineteenth-century geologist–explorer Clarence Dutton first entered the region he struggled for words to describe it. 'An army of caterpillars marching north' was his best effort. Most Americans, if they have seen this country at all, have probably done so from the air, and when seen from 30,000 feet caterpillars are not what comes to mind.

When flying west in the US I always book a window seat and ignore the in-flight movie so that I can revel in the desolate grandeur, and wonder at the secret workings of the Earth that the desert scenery unfolding below reveals. It's a landscape that has provoked profound thoughts, for study of its strange form helped Dutton to formulate his theory of isostasy and thus found the discipline of structural geology. Thanks to the work begun by Dutton, we now know that the army of mountain-sized caterpillars results from stretching and thinning of the continental crust, which at present is just twenty to thirty kilometres thick under Nevada—the thinnest under any continent. Some sections of the seriously weakened crust have dropped as many as three kilometres into the Earth, forming the region's basins, while other sections have risen 2.7 kilometres high to form the ranges.[13]

The crustal thinning was caused by one of the strangest movements in geological history. The Sierra and Cascade ranges swung westward as if on a titanic hinge fixed at their northern end, the southern parts of these ranges shifting up to 270 kilometres westward, stretching the crust under Nevada as they went. Most theories as to why this occurred depend upon a complex interpretation of movements between the Pacific and North American continental plates, resulting from the North American plate overriding the hot, crust-generating East Pacific Rise.[14]

During the Miocene volcanoes of the basalt-producing type joined those of the pre-existing explosive St Helens variety, and both now set to work altering the North American landscape. Flood basalts of enormous proportions spread over eastern Washington, Oregon, Idaho and northern Nevada, forming the Columbia River Plateau. In a matter of days the largest of the flows covered 40,000 square kilometres with molten rock heated to 1100 degrees Celsius. It seems possible that the same 'hot spot' in the Earth's crust that today fuels the geysers of Yellowstone fed these eruptions. If so, North America has slowly moved westward over the hot spot, so that the place where the heat appears at the surface has shifted east relative to the continent's margins.[15]

At this time an incipient rift in the Earth's crust—an embryonic African Rift Valley—opened on the eastern edge of the Colorado Plateau. Today this stillborn rift is occupied by the valley of the Rio Grande. The San Andreas Fault (which had begun life well out in the Pacific Ocean over 65 million years ago) then lay alongside the Pacific coast; it was busy rafting portions of Mexico northward at the terrific speed—at least for rocks—of one centimetre *per year*. The pivoting of various blocks carried north in this process opened enormous chasms in the Earth's crust, astonishing for their depth and narrowness. Downtown Los Angeles sits atop one such chasm. Its basement, consisting of rocks just 25 million years old, is buried beneath more than fifteen kilometres of more recent rocky debris. As this chaos geology was being concocted, Baja California was being ripped off the coast of Mexico, and would soon form its spectacular peninsula with its distinctive flora. By five million years ago these changes had created a continent whose outlines would be recognisable to a schoolchild as those of modern North America.

Tremendous regional diversity was occurring among plant communities at this time, and a marked north–south zone became established as seasonality and the harshness of the northern winter increased. Towards the end of the Miocene the first plant species unique to the Great Basin region came into existence, indicating that a true grassland steppe had developed there. In the Pacific Northwest, however, mesophytic forest—forests composed of trees that like conditions neither too wet nor too dry—still persisted.

Cactuses are unique to the New World, but no one is sure whether they originated in North or South America. Given their fleshy, segmented branches and the ability of some to put down roots wherever they touch ground, they have probably been dispersing over the sea between the continents for a long time. By the Miocene cactuses had become well established in North America and today they thrive as far north as the badlands of Alberta. By 10 million years ago the distinctive vegetation of the Sonoran and Mojave deserts, with their many cactuses, may have spread and acted as a barrier between other vegetation types to the east and west.[16]

The circumstances under which cactuses flourish long puzzled me, for some deserts have them in abundance while others lack them altogether. The cause eluded me until I met Professor David Baum, a botanist at Harvard University. Cactuses, he noted, store water in their expandable trunks and stems. They are in essence prickly vegetable camels (some species such as the cholla possess as many miserable qualities as their cantankerous mammalian counterparts). It may seem to make tremendous good sense to store water if one is living in a desert, but in reality this is not always the case. In deserts where rainfall is erratic there is a good chance that your reservoirs will not be replenished before they are exhausted. Thus in deserts where it might flood one year then experience no rain for three—such as those of Australia—cactuses are conspicuously lacking. Deserts that receive storms at predictable times of year (such as the Sonoran), however, are likely to be filled with vegetable camels.

One other factor that determines whether cactuses abound or not is the availability of ground water. They never have deep taproots, for in order to draw water from deep within the earth a plant needs thirsty leaves and relatively dry tissues. Cactuses have neither of these characteristics; by virtue of their very nature they cannot draw up this water. So it seems that desert plants face a choice: either to be cactus-like hoarders of water, or thirsty, deep-rooted plants like eucalypts; it is not possible to be both. The wisdom of pursuing either evolutionary path is determined by the kind of desert you live in. Erratic rainfall and deeply buried water resources will produce eucalypt-like species, seasonal rainfall and an absence of ground water encourage cactus-like species. Nutrient availability is also important here, for Australian deserts are notoriously nutrient deficient, and it is possible that those eucalypts with long taproots are searching for fresh, unleached rock, with its trace minerals, as much as they are hunting for water.

This understanding of the importance of seasonality for cactuses lets me view them in a very different light, for I now see them as just as much a product of North America's environmental determinism as the hibernation of squirrels or the fall colours of New England. All have their origins in the unerring seasonal procession, with its extreme swings of the climatic pendulum, that has been such a distinctive aspect of North America for the last 32 million years.

Talk of cactuses brings on thoughts of Gila monsters and rattlesnakes, and both make an appearance in North America at this time. The Gila monster and Mexican beaded lizard are the last survivors of an ancient lineage whose earliest representatives are found in 65-million-year-old European and North American rocks. This means that they, or their near relatives, were present in North America just prior to the asteroid impact, but did not survive the devastation there. It took them another 40 million years to belly-crawl back into the continent from their unlikely hideout in what is now France. The Gila monster (*Heloderma suspectum*)—that suspicious-looking pink-and-black-blotched salami of a lizard—has changed little since it arrived in North America over 20 million years ago. It is a once-extirpated then returned relic making a last stand in the deserts of the south-west.

Other venomous reptiles invading the continent at this time include the ancestors of the rattlesnakes (*Crotalus*) and copperhead (*Agkistrodon*), both members of the viper family, as well as the first front-fanged snakes (family Elapidae) which were similar to the coral snake (*Micrurus*). Rattlesnakes, of course, evolved in North America but their viper-like ancestors came from Eurasia. Non-venomous snakes also flourished in the Miocene, including the boas and the primitive aniliids. The sole survivor of this last family is the peculiar, burrowing, scarlet-and-black-banded pipe snake of northern South America, remarkable among living species for preserving traces of a pelvis.

By about 25 million years ago (the beginning of the Miocene) the first North American migration to South America occurred, with land tortoises making the ocean crossing. Some of these tortoises were very large, and relatives of these massive creatures have found refuge in the modern world on the Galapagos Islands. They were followed south soon after by the rear-fanged snakes (colubrids) and soft-shelled turtles. Then in quick order, between 23 million and about 12 million years ago, an extensive array of reptiles and amphibians rafted in the opposite direction—north from South America. As a result North America received its first whiptails (family Teiidae) since their extinction during the great catastrophe more than 40 million years earlier, the first anoles (*Anolis*) and frogs from the barking frog and narrow-mouthed frog families (Leptodactylidae and Microhylidae). All of these changes meant that, by the end of the Miocene, North America's herpetofauna was essentially modern.

Nine

GATEWAY TO THE PRESENT

The golden age of the Miocene ended five to six million years ago with the arrival of an unprecedented drought. The drought was a manifestation of a global climatic event known as the Messinian crisis, which was really just a continuation of the drying and warming trend that had characterised the entire period. Sediments underlying the Mediterranean Sea, however, indicate that conditions had become extreme.[1]

In the 1960s, scientists drilling into these sediments found huge amounts of salt that had accumulated on a dry seabed. Geological changes that narrowed the sea's opening, along with drought and the withdrawal of water from the world's oceans due to the growing Antarctic ice cap, were responsible. It seems that the swelling southern polar ice cap drove this cooling and drying phenomenon.

Most large species of North American mammals became extinct between the Messinian crisis and the onset of the current ice ages about 2.4 million years ago, with about thirty genera being lost in three million years. The victims included many of the truly native groups, including the protoceratids (horned relatives of camels), dromomerycids (an odd group of horned herbivores), the mylagaulids and eomyids (peculiar primitive rodents), and many horses, camels and pronghorns. This is also the time of 'the last rhinoceros in North America' as Cary Madden and Walter Dalquest titled their paper which describes a tooth found in 3.5-million-year-old sediments from Scurry County, Texas.[2]

The Messinian crisis also saw the final dismantling of the trans-American forests and woodlands, and is probably responsible for the rapid spread of grasslands worldwide. On the high plains and in the arid south-west, the woodland savanna and its marvellous mammal diversity gave way to prairie. The grass-covered Great Plains—a true steppe—came into existence only five million years ago and are still rather impoverished in species. Only with the onset of glacial times, 2.4 million years ago, did the steppe of Eurasia–Alaska come into contact via Beringia with the Great Plains, bringing with it new immigrants. By 1.7 million years ago this 'steppe bridge' had brought the mammoth and, by around 400,000 years ago, the bison. Before these and other Eurasian steppe fauna arrived, the North American steppe fauna was dominated by remnants of the native stock, such as horses, pronghorns and camels, whose roots go back 20 million years or more.[3]

Following the Messinian crisis North America experienced an increase in rainfall, possibly due to the rise of the Panamanian isthmus some 2.8 million years ago. This closed the portal that had linked the Pacific and Atlantic oceans, forcing warm water northward into middle and high latitudes along the east coast of North America. This warm seawater evaporated more readily than cold water, bringing more rain in summer and milder winters. At higher elevations and latitudes in North America the increased rainfall allowed snow to build up and ice to spread. The great northern ice cap—soon to be so influential in North American history—was on the move.[4]

The Pliocene period covers just 5.2 to 1.8 million years ago, but it allowed nature to put the finishing touches on some of the most spectacular scenery the North American continent has to offer. The key geological event in this regard was the uplift of the Colorado Plateau. This massive body of rock was swiftly raised about a kilometre and a half into the air, but it did not break up like the basin and range region. Instead it eroded into magnificence, producing the astounding geological features of Bryce and Zion national parks, the Natural Bridge National Monument and the Grand Canyon itself. This remarkable scenery is now so well known it needs little description, but the Grand Canyon is such a gargantuan feature that it never fails to provoke surprise. So huge is it, and so unexpected is the sight, that whenever I stand on its rim I find that my mind refuses to see it for what it is. Instead it somehow shrinks the scale of the phenomenon and I can't tell whether I'm looking at a toy canyon or the greatest gouge on the face of the Earth. That such a stupendous feature could be excavated over just a few million years speaks eloquently of the powerful geological and climatic forces that have shaped North America.[5]

As for the fauna, the most striking feature of the later Miocene and Pliocene was the scale of immigration from Asia. Two-thirds of all Pliocene mammal genera in North America are new appearances, and half of these arrived via Beringia. With the possible exception of the early Miocene, this constitutes the highest rate of immigration into the continent during the entire age of mammals.

One group of immigrants that quickly rose to dominance was the deer. The earliest North American deer appear about 5.2 million years ago, but do not become conspicuous until about 4.5 million years ago. One widespread kind of deer, however, may well be a true North American original, for the oldest well-dated caribou (or reindeer) fossils, which are around one million years old, have been discovered at Cape Deceit near Kotzebue Sound in Alaska. If their American origin is upheld, then the caribou/reindeer is unique among the larger animals of the Arctic tundra in having an American rather than Eurasian origin, which seems to be rather fitting given the enormous popularity that Hollywood's Rudolph has brought to the species. One other scenario, however, is that the reindeer first evolved in the circumpolar realm that rings the Arctic Ocean.[6]

The reindeer is a very strange and highly successful deer. It is unique in that both males and females have antlers, which are used to scrape snow from patches of lichen and other food in winter, and which may be used by the females to defend precious patches of food from males. And it is also the only deer to have been domesticated—the Sarmi people of far-northern Europe and Asia lead a semi-nomadic life, following their migrating herds and using gelded males to pull their sleds. It seems extraordinary that the people of North America, who have such a deficit of tameable creatures, did not also domesticate this animal.

Migration from North America to Eurasia was clearly on a smaller scale, although an impressive variety of animals made the crossing. Among the earliest were primitive wolves (*Eucyon*) and the ancestors of the Asian raccoon dog (*Nyctereutes*). These were the first members of the dog family to leave North America. Some North American shrews (of the tribe Blarinini) and hares also make the crossing, while true wolves (*Canis*), foxes (*Vulpes*), horses (*Equus*), elephant-like gomphotheres and some camels (*Paracamelus*) followed somewhat later.[7]

The migrations of the Pliocene provide an opportunity to summarise trends in the relationship between the faunas of North America and Eurasia in the 30 million years between *la grande coupure* and the onset of the ice age. As opposed to earlier times, when the balance of migration between North America and Eurasia was more equal, between 32 and 2.4 million years ago the balance became ever more lopsided. By the end of the period it was overwhelmingly from Eurasia into North America—the scale, diversity and repetition of the various mammalian groups involved in these invasions are awesome. Hundreds if not thousands of individual lineages are involved; they range from at least three separate invasions by elephants and their relatives, to hundreds by smaller mammals such as rodents and bats. The family that includes the weasels, skunks, badgers, otters and relatives (Mustelidae) typifies the pattern. Over the past 32 million years at least thirty-eight different genera of such creatures appear in the fossil record of North America, of which twenty-six are immigrants from Eurasia. The remaining twelve genera evolved on the continent from Eurasian stock, but not one single species is known to have migrated from North America to Eurasia.

In essence, for the past 32 million years North America has acted as a sort of Hotel California for the great majority of its inhabitants. It has been all too easy for them to come in through the Beringian front door, but has proved almost impossible for most to leave via that exit. Furthermore, the great majority of groups that established themselves in North America have in turn been overwhelmed by the new waves of immigrants continually arriving from Eurasia. This presents a striking parallel to the human history of North America, as will become apparent in acts four and five of this continental history. Nothing about the nature of Beringia can explain this enigma of biased immigration. Instead we must seek answers in the intrinsic nature of the North American and Eurasian faunas, and it is among those very few American success stories—those species that have colonised Eurasia—that the best clues can be found.

Incidentally, an analysis of non-mammalian vertebrates adds very little to our understanding, for Beringia has been all but inaccessible to most reptiles and amphibians, while among the birds only two or three American groups have established themselves in Asia, including the marsh wren (*Troglodytes troglodytes*) and the European buzzards (relatives of the red-tailed hawk and other buteos). How these very different birds became successful American invaders of the Old World while so many others failed remains a mystery.

As we have seen the dog family (Canidae) escaped the confines of North America about five million years ago after 35 million years of evolution in isolation. By three million years ago the true wolves and foxes had joined the exodus but, unlike other emigres, they returned repeatedly to their homeland, in the process becoming spectacular success stories and all but monopolising the cursorial-carnivore niche world-wide.

Camels left the continent around four million years ago after about 40 million years of evolution and found suitable ecological niches in the most forbidding of the Eurasian deserts. At least one subsequent invasion of camels from America occurred two million years ago. The third great group of American expatriates, the horses, has a more complex story. After 30 million years of evolution in isolation in North America they mounted their first invasion of Europe about 25 million years ago. Three separate invasions occurred between 25 and five million years ago, and then, about three million years ago the true (single-toed) horses migrated across Beringia. Despite the multiple invasions, through-out its history North America has remained the great forge within which the horse's evolution has been shaped. No Eurasian horses are known to have invaded North America south of the ice until 1492.

What, if anything, do these three groups of animals have in common? The two herbivore groups are both able to survive in marginal habitats. The camels thrive in deserts and on the highest, most barren mountains—places where life is truly precarious. Horses, because of their digestive system based on hind-gut fermentation, are capable of surviving on poor quality forage. It is thus no accident that Przewalski's horse (the last wild horse, named after its Polish discoverer) and the Bactrian (two-humped) camel found refuge in the same general region—the hostile steppe of northern Asia, for this is the kind of environment that such animals are built for.

The canids are somewhat different. They fill diverse ecological niches from top-order pursuit predator through to small omnivore, yet they too can survive in marginal habitats such as the high Arctic and the Kalahari deserts. I think that the larger canids have also evolved to cope with difficult environments, for they reproduce in a most distinctive way. These animals, from wolves to Cape hunting dogs, are social, and in each group only the dominant pair breed. The rest of the pack (usually genetically related to the dominant pair) forgo reproduction in order to help feed their young nephews and nieces. We must search as far afield as Australia's birds to find a similar social system of reproduction. Among these birds this phenomenon is called 'helpers at the nest': the 'helpers' refrain from breeding because the environment is so hostile that their chance of passing any genes into the next generation is improved by helping rear their nieces and nephews. It's a system well suited to poor environments for often two adults simply cannot supply enough food to meet the demands of the growing young. It seems to me that the peculiar canid system of reproduction may have evolved for pretty much the same reasons—it takes a pack to hunt with sufficient success to feed a litter, particularly when the adults are effectively 'tethered' to the den site. If this is true, then we can add canids to camels and horses as species that have adapted to life in extreme environments.[8]

North America has always been home to a great variety of mammalian species. Why is it that its three groups of world conquerors are all adapted to marginal environments rather than life in its richest regions? As we shall see, this pattern is not unique to North America. It seems rather to be a common property of most warm-blooded species that originate on smaller landmasses and invade larger ones, for almost all, although they may range widely, are adapted to life in harsh environments.

The three groups have been disproportionately important in human social evolution, for they include our oldest domesticate (the dog) along with what is arguably our most important animal acquisition, the horse.

Among the less widespread domesticates of North American origin are the two camels (Bactrian and Arabian), the llama and the guanaco.

Just why horses, camels and dogs have made such a huge contribution to our domestic menagerie is uncertain, though the adaptations they developed to survive in marginal habitats may have helped. In the case of dogs their unusual social system has clearly been an asset, for it has allowed them to slot into human social organisation. The link with horses and camels is less clear, but perhaps their adaptation to poor-quality fodder allowed them to survive the haphazard care doubtless provided during the early phases of domestication, or perhaps they could exist on the limited vegetation available within range of early human campsites. Humans too are a species that originated on a small continent (Africa) but went on to colonise a larger one. Is it possible we have formed such close relations with dogs, horses and camels because of some similarity in our deep evolutionary history?

UNITED LANDS OF AMERICA

Chimeras are freaks or sports of nature—organisms (sometimes even humans) that are made up of the cells of two different individuals that have somehow fused in early development. The New World is a curious mixture; a sort of chimera in biological parlance. If the chimera results from the fusion of, for example, a black and a white mouse embryo, it is relatively easy to tell which individual formed which part of the composite body because the result is a creature with a crazy patchwork of black and white skin.

We can think of the New World as a chimera because North and South America are fundamentally different landmasses that have fused to form a patchwork of organisms. While chimeric animals may look like a mad jigsaw puzzle they are not randomly assembled, for profound laws of development determine which cells migrate where as the body takes shape. This, I would contend, is also true of the New World patchwork, except here it is the laws of ecology and zoogeography that have determined the patterns.

As we have seen, North and South America were in contact just before the asteroid struck and it was then that possums and primitive placental mammals first entered the southern continent. They did not cross via the Isthmus of Panama (which did not then exist), but over a land bridge far to the east. Geologists have discovered that the Greater Antilles and a now-sunken land called the Aves Ridge (which lies just west of the Lesser Antilles) once formed a massive peninsula that they have dubbed GAARlandia, after an acronym of 'Greater Antilles and Aves Ridge'. This land was linked to South America but not to North, and existed between 50 and 30 million years ago. Earlier, between 75 and 65 million years ago, a more complete land bridge existed in the GAARlandia area that did link the continents. This proto-GAARlandia, made up of a volcanic arc, must have been a tenuous and short-lived structure but, as we have seen, it did permit a brief period of true continental interchange. Contact between the continents was severed when proto-GAARlandia began foundering some 65 million years ago, after which the plate bearing South America began a slow waltz to the south-west, out into the Pacific. The widening oceanic gap between the continents did not immediately become an insurmountable barrier, for adventuresome lizards like the South American iguanas reached North America by rafting. Nonetheless, it is fair to say that North and South America were just below each other's respective horizons for most of the age of mammals.[1]

The construction of the Panamanian land bridge, completed some 2.8 million years ago, involved the fate and eccentric movements of a number of ancient continental plates that constitute a true middle America. The largest and most significant of these is the Caribbean Plate, a piece of the Earth's crust that was jostled between its northern and southern neighbours until, some millions of years ago, continuous land emerged along its western margin, forming the region from Mexico to Panama. Some arcane geological studies suggest that this plate was formed in the eastern Pacific Ocean and that it drifted over 2200 kilometres westward to come to rest in its present position. Other geologists dismiss this view, disputing whether the Caribbean Plate is even a single entity, citing evidence that the plate itself consists of three separate and unrelated geological units. The northernmost of these (which forms southern Mexico to the Yucatan Peninsula), they argue, has been part of North America since the middle of the age of dinosaurs some 100 million years ago. The central section (what is today Honduras), they suggest, swung into its present position from a region off the west coast of Mexico by the middle of the age of mammals, some 30 million years ago. The origin of the southernmost section (Nicaragua to Panama) is unclear, but it was in place by about three million years ago.[2]

Biologists have since weighed into the argument, suggesting that the peculiar insectivores of the Caribbean, such as the solenodons, have their closest relatives in far-off Africa, indicating an Atlantic origin for the plate! It seems that we are far from having a complete understanding of the geological history of this complex region, but what is certain is that the land bridge that formed was to have a profound influence on both of its larger neighbours.

We know that the size of a landmass is an important factor in determining its inhabitants' fate when other landmasses come into contact. North America is approximately 25 per cent larger than South America, and its intermittent connection to Eurasia gave it a much larger 'virtual' size than it would otherwise have had, by filling it with immigrants from a truly vast continent. As a result, since the Pliocene (five to two million years ago) the diversity of the North American mammal fauna has exceeded that of South America by 60 per cent. One would expect, given these factors, that the Panamanian land bridge heralded a northern takeover of South America, but this is not quite what happened, for biology is complex and many factors come into play when worlds collide.[3]

The great biological interchange began well before the Panamanian land bridge finally emerged, for by 10 million years ago various organisms were crossing the narrowing water gap between the continents, finding new land and exploiting its resources. Among the earliest of such immigrants were the birds, and this interchange proved to be a dramatic victory for the south, for it resulted in a North American avifauna with a large component that is South American in origin. Some of the most familiar of these are the New World tyrant flycatchers (which arrived so long ago that they have given rise to a spectacular radiation of North American species) the vireos, the paruline warblers and the New World vultures. Curiously, almost all of these South American elements share one characteristic in common—they are migratory land birds.

The migrations undertaken by North American birds are a marvel—the sheer number of species and individuals flying south astonishes the observer. The flocks follow four major continental flyways of which the Mississippi River marks the most important route. Gliding species such as the birds of prey prefer to follow the uplift created by the Appalachians, while it has been recently discovered that many species, including tiny perching birds, take an oceanic route which carries them far out over the Atlantic before they arrive at their destination.

Many people tend to view migrating birds in an anthropocentric way, imagining that they are going south for a holiday. The truth for the migratory land birds, however, is that most are flying home, to secure their vital ecological niche space in the crowded tropical rainforests of Central and South America. These birds are called northwards to breed by the great North American climatic trumpet. It heralds in a New England summer which is Puerto Rican in temperature and probably exceeds it in abundance of insect life, yet the deadly northern winter it brings in turn is so harsh that it reduces the resident avifauna to a mere 'skeleton crew' of species. The South American migrants arrive in the north at a time of abundant food, for the overwintering birds cannot breed up fast enough, nor are they diverse enough, to fill the niche space that opens in the northern forests during summer. The migrating species fatten their chicks in what must seem to them a truly promised land, where food is there for the taking and where, for all they see of it, summer reigns eternal.

At first it may seem strange that North American birds failed to take advantage of seasonal differences in South America. North American species attempting to do this, however, are faced with a much more difficult task than are South American species migrating north. Any North American migrants arriving in the southern rainforests would find an ecosystem packed with hundreds of species of birds, each occupying a narrow and defined ecological niche. No generalist or newly arrived species could hope to compete against such specialists. These forests would in effect put out the 'house full' sign to any new species hoping to find a winter refuge there.

The North American migrating waterbirds such as geese and ducks seem to have a different origin and to face other difficulties. Many belong to groups long established in the temperate regions, and their migrations can be imagined as wanderings in search of water—abundant in the north in summer, but only found further south in winter, after the North American climatic trumpet has turned the northern lakes and rivers to ice.[4]

The patterns of survival and immigration that we have seen so far have led to the development of a richly mixed North American fauna. Its mammals are overwhelmingly Eurasian in origin, its migratory land birds largely South American, while much of its herpetofauna is unique. No other continent exhibits such different origins for the constituent parts of its fauna, which nevertheless reflects so accurately the zoo-geographic history of the host landmass.[5]

It was not only birds that took early advantage of the narrowing water gap between North and South America. Plants—including those unique

New World creations, the cactuses—must also have crossed early and often. The cactuses are well served by specialist pollinating bees; their showy flowers often evolved in tandem with individual bee species. For some reason that has not yet been divined, specialised, cactus-pollinating bees are absent within ten degrees of the equator. Thus these insects are one group that has not crossed the Isthmus of Darien. This explains why bees pollinating cactuses to the north of this zone are very different from those living to the south. Most of the northern bees are derived from ancient temperate North American lineages, while those of South America are from Gondwanan stock, having more in common with Australia's bees than those pollinating similar cactuses in North America.

Some mammalian drifters also crossed the Panamanian sea long before a land bridge formed and, curiously, South American mammals reached North America long before any northern species managed to push south. The first to make the journey were the South American ground sloths. These enormous shaggy herbivores were, to judge from their living relatives the arboreal sloths, excellent swimmers with very low metabolic rates. Both of these characteristics, along with their large size, would have helped them enormously in making the crossing. Their low metabolism would have allowed them to 'shut down' to some extent when washed out to sea, permitting them to endure a drop in body temperature and an absence of food and fresh water until currents and their own swimming brought them to land. Such advantages saw representatives of two separate families of giant ground sloths (the Megalonychidae and the Mylodontidae) make the crossing by eight million years ago.

The first North American mammals to travel south were members of the raccoon family. They arrived around four million years ago, and once ensconced in South America evolved into bear-sized and bear-like creatures, something they had never done in their old home. No sloth immigrant ever evolved in such a radical way, and we shall see more of this pattern in the subsequent history of exchange. The living South American members of the raccoon family, incidentally, are not related to these giants but to much later immigrants.

Not surprisingly, the raccoons appear to have been followed by those masters of dispersal, the mice. Their date of arrival in South America is unclear, although they seem to have been present by four million years ago. It is also uncertain whether one or a number of lineages arrived, but soon after arrival they began one of the most extraordinary and explosive evolutionary events ever witnessed on Earth. Today the mouse and rat-like sigmodontine rodents are represented by over 200 genera and 424 South American species. No one knows how many there are for sure (the scientists are still counting) and every year brings discoveries of hitherto unknown species and even genera. So diverse are these mice that one out of every four of all South America's mammal species is a sigmodontine rodent! To see comparable diversity created in such a short time one has to return to those days of vacuum-driven evolution following the end of the age of dinosaurs, when the entire blasted planet was inherited by a few rat-sized mammals.

By 2.8 million years ago the first dry land connections to exist between North and South America for at least 62 million years had been established. Then the Isthmus of Panama was a flat, probably seasonally dry plain covered in woodland and savanna. The fauna of North America had been adapting to drier, more open conditions for millions of years, and this land bridge was a veritable highway to them. The charge south over the open plain was led by the skunks and peccaries. Horses quickly followed, then came a mass invasion of dogs and foxes, bears, weasels and their relatives, cats great and small, squirrels, shrews, rabbits, tapirs, camels, deer and gomphotheres. These creatures were to transform South America, sweeping away many archaic groups and fixing in their place a fauna of predominantly North American origin. As a result of this invasion, South America today has more kinds of dogs and their relatives (a truly North American group) than any other continent. It is home to over half of the world's camelids (the vicuna, llama and guanaco—also true North Americans), the last tremarctine bears (relatives of the formidable short-faced bears), three-quarters of the world's tapir species, as well as a stunning range of cats, deer and peccaries. Indeed, following the extinctions that carried off most of the old North American groups 13,200 years ago, South America was destined to become the last refuge of many ancient northerners. Today, half of the modern South American mammal genera are derived from North American groups that arrived during the great interchange and diversified there in less than three million years.[6]

The success of the woodland and savanna-adapted mammals is partly explained by the fact that temperate North America (where woodlands and savanna dominate) has an area of 21.2 million square kilometres north of the Isthmus of Tehuantepec, whereas temperate South America (Chile, Argentina, Uruguay) has an area of just 3.7 million square kilometres. The grassland and woodland habitats of the early land bridge allowed the faunas of these two regions to come into intimate contact, with the size of their original habitat explaining the success of the North American mammal immigrants when they arrived in South America. It cannot, however, explain the extraordinary radiation of many species once they reached the south.

A few species of birds joined the mammals in their southern voyage, among them two kinds of jay, whose arrival heralded the completion of a long journey by the crow family which began in Australia over 30 million years earlier. Reptiles also crossed during this period and South America got snapping turtles, the first ever to have left North America since they evolved there in the age of dinosaurs. Then just two million years ago, in a final gift from the north, rattlesnakes crawled south across the isthmus.[7]

A few kinds of Latin mammals did manage to push their way north against the invading throngs. They included the porcupines (*Erethizon*), the large rodent capybaras (*Neochoerus*), the armoured glyptodonts (*Glyptotherium*), and more ground sloths, this time of the genus *Glossotherium*. These invaders from the south were later joined by two kinds of armadillos (*Dasypus* and *Kraglievichia*) and yet another giant ground sloth (*Nothrotheriops*).

By two to 1.9 million years ago, more capybaras (*Hydrochoerus*) and creatures resembling giant armadillos called pampatheres (*Holmesina*) had crossed. By 1.4 million years ago a llama (*Palaeolama*, itself evolved from North American immigrants that had made the journey south a million years before), the ancestor of the opossum (genus *Didelphis*) and—you guessed it!—more ground sloths (giants of the genus *Eremotherium*) had joined them. By a million years ago anteaters related to the living giant anteater had reached as far north as what is now Sonora in Mexico.[8]

The strangest of all the creatures to travel north at this time was an enormous, flightless bird belonging to a family that had dominated the large carnivore niche in South America for over 50 million years. Three metres high, 400 kilograms in weight and with a beak thirty centimetres long and shaped like a hatchet, these birds known as phorusrhacoids were formidable predators. One of their strangest features was their wings, for they had re-evolved an opposable 'thumb', which was evidently used to grasp struggling prey. The species that crossed into North America, known as *Titanis*, was one of the largest of its kind. Its distribution in North America was always limited to the south, principally in Florida, where its fossilised bones remained undiscovered until 1963. Not since the formidable *Diatryma* stalked the North American forests 50 million years earlier had the continent seen such avian predators, and clearly *Titanis* was a capable hunter, having to compete with advanced mammalian predators such as canids and felids.[9]

I have often wondered whether the development, on two separate occasions (once each on North and South America), of such great avian predators tells us something about the determinants of evolution in the New World. In other words, do North and South America share something that encourages the evolution of birds like *Diatryma* and *Titanis*? For it is these two continents, and these alone, that have given rise to such monstrosities and they have arisen, it seems, from two entirely different origins—storks or screamers in the case of *Titanis*, and rails for *Diatryma*. Some as yet undiscovered ecological principle may explain this occurrence, but upon reflection I can only put it down to coincidence. Yet it is a coincidence as great and seemingly portentous as that which resulted in the sun and the moon appearing to be the same size in our skies. It pays to remind ourselves that not all in nature is meaningful and informative of the evolutionary laws by which we live.

One vast migration still to be mentioned involved an entire biota. This invasion must have happened once wetter conditions were established over the Panamanian land bridge in the last one or two million years, for it involves rainforests. Very few plant groups inhabiting the Central American rainforests have North American roots—vines of the Dutchman's pipe and grape families being two rare exceptions. Even the pollinating insects of the Central American forests are largely South American in origin. The freshwater fish of Central America show a similar pattern with all but seven species hailing from South America. These exceptions—a garfish (*Atractosteus*), three clupeids (*Dorosoma*), a catostomid (*Ictiobus*) and an ictalurid (*Ictalurus*)—constitute fewer than 5 per cent of the total Central American freshwater fish species. Similar patterns of South American dominance are seen in almost all other groups of organisms inhabiting these Central American rainforests.[10]

This astonishing resemblance of the Central American rainforests to those of South America is entirely due to the northern migration, *en masse,* of a complete southern rainforest biota into a region which, whatever its origins, was firmly attached to North America at the time of first contact between the continents. It is as if Central America swapped horses halfway through a race, ditching its North American equids and other fauna and welcoming a wholesale migration from the south.

Given the tendency of species from larger landmasses to triumph numerically during periods of faunal interchange, the success of the rainforests of Central America seems anomalous, for they are populated by organisms from the smaller southern continent. The answer to this paradox lies in the shape and history of climatic zones in North and South America. The percentage of North America lying in the tropics has always been small, and more northerly regions that may have acted as subtropical extensions of the torrid zone have, for the last 32 million years, been subject to punishing North American winters generated by the great climatic trumpet. Thus the truly tropically adapted fauna of North America can be thought of as occupying a small 'island' in the south. The very rules that determine the fates of mice and men would ensure that they would be overwhelmed when they came in contact with the vast tropical provinces of South America.

A few later southern immigrants into North America are worthy of closer consideration. At a date unspecified, yet very close to our own age, a notoungulate known as *Mixotoxodon* colonised Central America, probably as part of the southern derived rainforest assemblage. These creatures, looking rather like odd hippopotamuses, were the last of the lineage that had so confused the American Museum of Natural History's Mr Matthew in 1913, prompting him to write to poor Stein about a possible mix-up in his specimens from Wyoming. This time there was no doubt about it; the notoungulates had finally arrived, 57 million years after they were first (and falsely) ushered onto the North American stage.

Then, around 10,000 to 40,000 years ago a final, late-coming southern plant immigrant arrived in North America, if not on the wings of a dove then perhaps in the tail feathers of a plover, for this is how scientists think that a few seeds of the creosote bush (*Larrea tridentata*) made it to the American west, all the way from Argentina. This strange, sticky bush now dominates the desert flats of the south-west, which is surprising given the overall lack of success of other temperate South American invaders in North America. Its success is due in part to its tenacious competitiveness—the creosote's roots starve grasses and release toxins that kill its woody neighbours. Just why these characteristics were selected for in an Argentinian temperate plant and not a North American one is unclear, but another factor for its success must have been its ability to leave its predators and parasites behind in Argentina, for they could not travel on the tail feathers of a plover. This meant that, like a eucalypt growing in California, the plant could flourish untroubled by the insects that burden its stay-at-home relatives.

Had the creosote bush arrived last century rather than 10,000 years ago, it would doubtless be proclaimed the most noxious weed ever to have invaded North America. Yet ten millennia have been long enough for moths and stick insects to develop into endemic species that depend solely upon the creosote bush for a living. Such species are perhaps slowly robbing the creosote bush of its advantage, and in another ten millennia it may well be just another moderately successful bush in the diverse deserts of the American south-west.[11]

There is a curious codicil to this tale of intercontinental invasion. As we have seen with llamas, some immigrant groups, having evolved into distinctive South American forms, mounted a successful re-invasion of North America from the south. The peccaries seem to be one such group. They are an ancient part of the North American mammal fauna, being placed in their own family the Tayassuidae, (formerly known as the Dicotylidae meaning 'two belly buttons', in reference to a navel-like scent gland on their backs). These pig-like creatures arrived in North America from Asia about 33 million years ago and invaded South America 2.8 million years ago. Remarkably, all of the current North American species appear to be derived from South American ancestors which re-invaded the north in the last few hundred thousand years. The collared peccary is an extreme example, for it may have reached North America less than 10,000 years ago. Just why the North American lineages became extinct, leaving only South American ones to inherit North America, is not yet known.

The great faunal exchange initiated by the Panamanian land bridge was described by the famous palaeontologist George Gaylord Simpson as 'one of the most extraordinary events in the whole history of life'. It was a vast natural experiment; a promiscuous mixing of whole continents full of organisms that had been isolated for tens of millions of years. Some aspects of the interchange, as we have seen, abided by the laws of zoogeography in that the inhabitants of larger landmasses (or larger climatic zones within those landmasses) dominated. Some phenomena generated by this vast interchange, such as the failure of the cactus-pollinating bees to cross the equator, remain mysterious, while others seem to be due to the effects of secondary laws. Two of the most outstanding such dilemmas concern the propensity of the North American invaders to have undergone explosive evolutionary radiations while the South American invaders by and large did not, and the success of seemingly unpromising South American invaders such as the sloths.[12]

The spectacular diversification of groups like the sigmodontine rodents and canids can be explained in the same terms as the diversification of the placentals after the asteroid impact. Such evolution just has to be vacuum-driven, though in both cases the factors causing the ecological vacuum were different. Sixty-five million years ago the vacuum had been caused by extinction, while around three million years ago in South America it appears to have been caused by a paucity of evolutionary activity—there was nothing like a mouse or a fox on the continent at the time the first sigmodontine rodents and canids arrived. To them South America looked like an open field, and they responded accordingly. Just why such creatures had not evolved in South America is unclear, but it may be related to the relative sizes of the continents and the initial, limited fauna present when South America was first isolated. It is also worth remembering that South America had no carnivorous placental mammals before the great exchange, for its carnivores were marsupials, the larger kinds of which became extinct around the time of contact. Thus dogs at least (or anything like them) could not evolve in South America. Mouse-like rodents were also absent, as they were in Australia until about four million years ago. In Australia today, as in South America, mouse- and rat-like rodents comprise 25 per cent of the fauna, providing a remarkable parallel of events.

What do the few successful South American mammal immigrants into North America tell us about the nature of the mammals that swim against the tide—those that travel from small continents to large? Why, one feels compelled to ask, should those ponderous archaic-looking beasts, the sloths, be so immoderately successful as to have invaded North America on five separate occasions? A glance at the living sloths hardly leaves one impressed with their capabilities. A biologist friend of mine who admires leaf-eaters of any sort has called them 'virtuous, dignified and restrained'. A less kind critic might characterise them as primitive and slow-witted, with appallingly low metabolic and reproductive rates. Their very name suggests that they should not have been successful competitors against the diverse and vigorous North American fauna. On top of that, armadillos and their relatives—yet another group of rather primitive mammals—contributed another three successful immigrant groups while that archetype of stupidity the opossum also managed to make a fist of it in the competitive North American environment. This seems bizarre given that North America had purged itself of marsupials at least 18 million years earlier.

The success of these creatures says something about an important albeit secondary law influencing the outcomes of faunal interchanges. We saw earlier how the successful North American immigrants into Eurasia were those adapted to harsh, often marginal environments: camels in their deserts; horses, canids and cheetahs on their plains. Back immigrants (as I have dubbed species that successfully migrate from a small land to a larger one), it seems, tend to slot themselves into niches where there is little competition and, for large mammals, these are often at the nutrient-poor end of the environmental spectrum. These species occupy environments so difficult that they cannot support the locals in the manner to which they have grown accustomed.

Ground sloths are probably the prime example of this. Although extinct, it is apparent from their remains and their living relatives that they could survive on very poor-quality browse. No North American endemics could quite match their ability to live off the smell of an oily rag, and so the South Americans survived, but did not diversify from their niche. For all their success these creatures had an Achilles heel. It pays to be large if you must live on such a miserable diet, yet these great creatures could not afford the metabolic luxury of the best defence for large creatures in such a changing world—rapid flight from danger. They became cumbrous, dependent on bony armour and great claws to defend themselves.

The lack of diversification of these lineages in North America may also be attributable to the ecological niche they occupy, for poor or otherwise limited environments can support only limited populations of large, warm-blooded creatures. Such environments cannot support the species diversity of richer regions, as is seen in the contrast between the spectacular antelope diversity of east Africa versus the lower diversity of the north African deserts. A sloth surviving in North America may have found that it needed to remain a generalist, occupying a broad ecological niche, which again may have militated against speciation.

The success of the opossums may have been due to their adoption of a different evolutionary strategy, for they have become the ultimate throwaway mammal. They seem to eat almost anything regardless of palatability or toxicity, they breed quickly and their lives in the wild are brief, two years or less. They have developed a similar strategy to weeds, for they thrive in disturbed habitats. The dramatic disturbances of the ice ages and perhaps of Indian agriculture may have aided them, and the European invasion was certainly a godsend, helping them spread north of the Hudson in the 1840s and (through human-aided trans-location) to invade California in the twentieth century.

Water-crossing wayfarers from the south have been arriving in North America ever since the first great ground sloth hauled itself ashore on the continent over eight million years ago, and almost all migrants from the south share a great commonality of experience. Virtually all, I think, existed on minuscule energy budgets in their new homes until a few lucky ones (like the opossum) found a new but dangerous way of living. They have had to subsist on poor-quality food in niches that their privileged North American counterparts did not deign to fill.

ELEVEN

LAURENTIDE

I write these words from an institution that has a special place in the study of ice ages. It was founded under a bequest that states that 'neither the collections nor any building which may contain the same shall ever be designated by any other name than the Museum of Comparative Zoology at Harvard College'. Yet so influential was its founder that it is known the world over as the Agassiz Museum.

Louis Agassiz was an extraordinary man remembered for many things, but of all his achievements perhaps the most celebrated was his 'discovery' of the ice age. Agassiz was a Swiss naturalist who, during a tour of the Rhone Valley in 1837 in company with a local geologist, realised that many typical glacial features could be seen throughout Europe in areas which are today remote from ice. His book *Etudes sur les Glaciers*, published in 1840, outlined the then radical idea that in the not too distant past much of Europe had lain in the grip of glaciers. When Agassiz came to Boston in late 1846 he revolutionised the natural sciences in North America—and found plenty of evidence for his beloved ice age.

In New England this last is not a difficult thing to do, for only 20,000 years ago North America was a veritable empire of ice.[1]

The causes of the onset of the ice age some 2.4 million years ago are still fiercely debated, but are probably to be found in a continuation of the same forces, tectonic and other, that induced previous cooling events. These include alterations in oceanic currents, tectonic plate and volcanic activity. Ice ages are composed of many cycles of ice advance and retreat. Periods of advance are known as glacial maxima while the retreats are called interglacial periods. At present, we are experiencing an interglacial period and the world is about as warm as it gets in an ice age.

Small variations in the tilt of the Earth on its axis and variations in the planet's elliptical path around the sun are all that is necessary to plunge the planet in and out of the freezer. Just 10,000 years ago, when it was rapidly warming, the Earth came closest to the sun in July and as a result the northern hemisphere got 8 per cent more radiation in summer than it does today. This proximity contributed to the melting of the northern ice cap, and indicates that apparently minor events can produce significant changes in ice volume. The celestial changes that cause oscillations between glacial maxima and interglacial periods are known as the Milankovich cycle, and they throw the switch from a glacial maximum to an interglacial period about every 105,000 years.

During an ice age it's as if the Earth is balanced on a fine knife-edge—so fine indeed that the tiny changes of the Milankovich cycle, which go unnoticed during warmer times, cause massive climatic shifts. On one side of the knife-edge lie times of relative warmth like the present, while on the other lies frigidity—an icehouse world. Worse, there is no middle ground, for we flip-flop from one state to the other with just a slight push from the sun—and the flip-flops can be incredibly rapid. Some scientists even argue that the shift from one state to another occurs in as little as three to five years, although of course it takes far longer (thousands of years) for the ice to melt or build to its maximum extent.

Palaeoclimatologists tell us that another glacial maximum is due 'any day now' on the geological time scale, but this scenario of pending climate change ignores the wildcard played by greenhouse emissions. Strong arguments have been made that they will result in a warmer world and equally strong ones have been put that greenhouse gases will hasten the onset of the next advance of the ice. At present frightening uncertainty stares us in the face.

Over the past 2.4 million years the Earth has flip-flopped in and out of the freezer at least seventeen times. The last freeze, known as the 'last glacial maximum', grew rapidly in intensity 35,000 years ago, peaked 18,000 years ago and deteriorated from about 15,000 years ago. On land, each ice age tends to obliterate signs of its predecessor, so measuring the severity of the various advances of the ice can be difficult. The sea preserves a better if less accessible record, and its sediments suggest that during the penultimate glacial maximum 150,000 years ago the sub-Arctic front of the northern oceans stood five degrees closer to the equator than it did 18,000 years ago. It may have stood even further south on several occasions before that.[2]

Because the last glacial maximum left such recent and well-preserved evidence we tend to use it as a yardstick of what the world is like in the grip of such cold. To study the rocks of that age is a disconcerting experience. The scientists who have done so have discovered that 18,000 years ago a mind-numbing 80 million cubic kilometres of water was frozen in glaciers, lowering sea levels by around 120 metres. The seas would have dropped further had not the great weight of ice pushed down on the world's continents, deforming the crust in such a way as to raise the ocean bed, displacing water upward. At this time half of the world's oceans were choked with icebergs, while Beringia, that link between Asia and North America, was a broad plain 1600 kilometres wide from north to south.

North America in its inimitable manner amplified this great freeze, for while global temperature dropped by an average of four to five degrees Celsius, temperatures in the American mid-west plunged by an estimated eight to ten degrees Celsius. Britain, in comparison, cooled by around six degrees Celsius. The world's oceans also cooled differentially, the North Atlantic becoming much colder than the North Pacific because the Bering land bridge cut off access between the warm Pacific and refrigerated Arctic oceans.

It is the issue of ice, however, that really beggars belief when defining the North America of 18,000 years ago, for the frozen facts are astonishing. Today we think of Antarctica, with its 12.6 million square kilometres of glaciers (out of the world's 14.9 million) as the home of ice. North America is a veritable Riviera in comparison, for it supports less than 77,000 square kilometres of ice, just 0.7 per cent of that presently encrusting Antarctica.

Eighteen thousand years ago, though, North America eclipsed Antarctica in the ice-accumulation stakes. Then Antarctica supported only slightly more ice (13.8 as opposed to 12.6 million square kilometres) than it does today, but one North American ice sheet alone—the awesome Laurentide—sprawled over 13.4 million square kilometres of land. That one sheet was more extensive than all the present-day glaciers of Antarctica put together. But the Laurentide was just one of North America's ice sheets. Others, including the Alaskan and Canadian Cordilleran ice sheets, were also of massive dimensions. If we include the Greenland ice sheet (which probably contacted the Laurentide), then in all some 18.5 million square kilometres of glacier encrusted North America.[3]

These colossal figures tend to numb the mind, but comparisons with the situation elsewhere on the globe can help us comprehend their significance. All of Eurasia for example—a landmass over twice the size of North America—supported no more than 11 million square kilometres of ice during the last glacial maximum. South America supported just

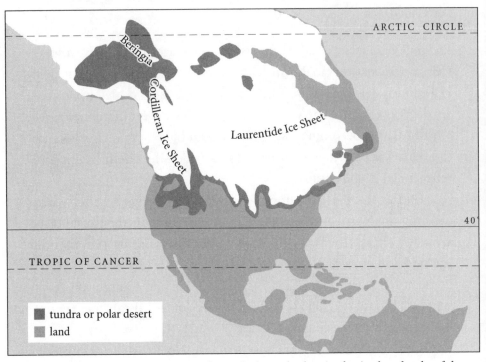

Beringia

Cordilleran Ice Sheet

Laurentide Ice Sheet

40°

TROPIC OF CANCER

■ tundra or polar desert
■ land

The empire of ice: 18,000 years ago North America lay under the grip of an ice sheet that dwarfed that of present-day Antarctica.

870,000 square kilometres, Africa just 1900 square kilometres and mainland Australia a pathetic 50 square kilometres. Since the ice began to retreat North America also experienced the greatest thaw. It now supports 76,880 square kilometres of glacier (0.4 per cent of its original amount), while Eurasia still supports about 125,000 square kilometres of ice sheet (1.3 per cent). The other continents are hardly worth mentioning for the amount of ice they support today.[4]

Researchers have tried to calculate the volume of water bound up in the various ice sheets. Such studies indicate that the Laurentide ice sheet alone contained enough water to lower the level of the sea worldwide by up to seventy-four metres, there being about 27 million cubic kilometres of water frozen within it. All the Antarctic ice sheets put together at the height of the ice age contained 26 million cubic kilometres of water, while the European glaciers (the next largest) contained just

12 million cubic kilometres. There is no more splendid example of North America's capacity to act as a great amplifier than this: the Earth cools by an average of four to five degrees Celsius, but America's heartland chills by around ten degrees Celsius, and North America overtakes Antarctica as the great accumulator of ice. If any nations have special cause to fear global climate change, it is surely those that call this great climatic amplifier home.

What is North America like during a glacial maximum? Ice sheets over three kilometres high at their centre cover most land north of New York. There are a few polar oases in areas that receive little precipitation, the most important being in Alaska and Beringia. The weight of the ice actually deforms the continent, bowing the rocks until they sit about 370 metres below their present level—so extreme is this effect that the continent is still rising following the melting of 12,000 years ago.[5]

In a glacial maximum the seas drop by around 120 metres, exposing a great coastal plain right round the continental margin, opening up or enhancing connections with Eurasia, South America and Greenland. The Atlantic is frigid and berg-choked while the Pacific remains quite warm. The ice itself forces great changes, gouging at mountain ranges and rafting rocks up to four kilometres long—only slightly shorter than Australia's Uluru—hundreds of kilometres from their place of origin.

To the south of the ice stretches the North American tundra belt, a habitat seen at its best today around Hudson Bay and regions north. Eighteen millennia ago tundra grew on Manhattan, reindeer frolicked in New Jersey, bison roamed Florida and, in what is now the arid southwest of the United States, forests grew around large freshwater lakes. This was due to radically different weather patterns caused by the huge ice mass sitting over the north of the continent, which split the westerly jet stream into northern and southern sections.

The jet stream marks the boundary between warm and cold air masses. Today it crosses the Pacific coast at about 50 degrees north (the latitude of Vancouver), turns south-east then crosses the east coast at 40 degrees north (close to Fort Monmouth). Eighteen thousand years ago the northern branch flowed across Canada to exit in New England, while the southern branch flowed over New Mexico and Arizona. The jet stream controls the location and path of storms, and 18,000 years ago it brought ample rains to what is now desert.[6]

Plant distribution and to some extent vegetation communities were quite different 18,000 years ago from what they are today. Then, one of the most important plant communities in the eastern US was a parkland of spruce (*Picea*), which covered a large band of country to the south of the ice. There, the trees must have grown far enough apart to allow for vigorous growth of herbs, for vast quantities of pollen have been preserved in bogs and lakes. The spruce parkland persisted until 14,000 years ago, but as the Earth warmed it disappeared, vanishing completely by 12,000 years ago.[7]

To the south of the spruce grew northern pines like the jack pine, which covered territory from Georgia to the Carolinas. Studies undertaken in Camel Lake on the gulf coast of Florida reveal that around 14,000 to 16,000 years ago a plant community rich in pines and spruce thrived even that far south. This community has its nearest living analog in the plants growing today in the Montreal region of Canada. If even northern Florida was occupied by such cold-adapted forest, where, one wonders, were the deciduous forests that so predominate in the eastern parts of the US today? To find evidence of them we must go south in Florida to Sheelar Lake which lies near the centre of the peninsula at its northern end. There, in sediments that accumulated 16,000 years ago, we find evidence of the broadleaf deciduous forest that, when the ice melted, would inherit half a continent.[8] This forest, which reaches its apogee today in New England, thus found a refuge only on the very south-eastern tip of the US. One wonders what America would be like today if Florida was a little shorter.

During the glacial maximum, vegetation that today grows on the summits of high mountains would have extended into the lowlands. In the south-west of the continent ponderosa pine parkland extended into areas that today support cactus. It would have been fringed by pinyon–juniper woodland that grew where middle elevation vegetation now flourishes. Then, desert plants such as the cactuses probably grew only in the lowest parts of the Sonoran Desert in the US and may well have been banished south of the border entirely, finding a final refuge in Mexico. Squeezed between these vegetation types were sagebrush and chaparral.[9]

The Great Plains had an entirely different character at this time too. The influential palaeontologist Paul Martin wrote:

> I do not claim that prairie vanished completely 20,000 years ago, but my view through the crystal ball is certainly clouded by conifer pollen. South of the glacial ice covering Illinois and Iowa, and beyond a narrow tundra belt, I envision a region of conifer forest or parkland and savannas extending into central Texas and Louisiana. To survive, the prairie herbs had to creep under the trees. The vast, treeless prairie of historic times from Manitoba to the Gulf of Mexico cannot be traced back in the fossil record beyond the past 11,000 years.[10]

The shift from forest to prairie on the Great Plains may have had as much to do with the coming of humans as climate change, for there is evidence that Indian fire played a role in keeping this vast grassland tree-free. Thomas Jefferson was the first to suspect as much and his theory has received support from recent researchers.[11]

Despite the vagaries of the ice age there were some sanctuaries of stability. The west provided a last redoubt for the Douglas fir; the Sierras for Sierra redwoods and other, smaller plant species. Prior to the ice age some such plants had a far wider distribution than the present-day: hemlock and Douglas fir were found as far afield as Europe until a few hundred thousand years ago. Then the advance of the ice backed them up against the east–west running European Alps and, unable to find a way south through the barrier as the climate cooled, both became extinct there. In the American west, however, the north–south orientation of the mountains provided the tall trees with a migration highway south, while the Pacific Ocean ensured abundant moisture and some moderation of climatic extremes. The timber-poor Europeans have always coveted tall, straight trees and after the loss of the American colonies in 1776, Britain found herself short of such timber for masts. The return of the tall trees was only a matter of time and by 1827 the Douglas fir—newly transported across the Atlantic—was once again growing in Europe.

TWELVE

VISIT TO A NEW WORLD

What would we have seen if we could have visited the American west of 18,000 years ago? In geological terms this is a very brief time—too short for any new species of larger creatures or even plants to evolve. Thus the vegetation would have been immediately recognisable, even if it was growing in strange places and in strange associations.

Living among these plants, we would have found a fabulous array of large creatures. If the abundance of elephants in Africa today is any guide—where they can exist at a density of three per square kilometre—the most visible animals among the ice-age plants would have been the mammoths and mastodons. The mastodons were the last survivors of an African invasion that began 17 million years earlier, an echo of the pseudo-African fauna of North America's golden age. The ancestors of the Columbian mammoth (*Mammuthus columbi*) had arrived on the continent from Asia just 1.7 million years ago. They settled in the west of the continent to the south of the ice. Later the famous woolly mammoth, which inhabited the Siberian and Beringian steppe, arrived in Alaska.

Ice-age art gives us an excellent idea of what the woolly mammoth looked like. Their high shoulders, humped head, sloping back and low hindquarters make them very different in appearance from the living elephants. We also know from frozen remains that they had tiny ears, their trunk-tips had broad flanges running up their sides, their bodies were covered in dense black hair and under the skin they laid down an insulating layer of fat around seven centimetres thick.

Although we have no mummies or ice-age illustrations of the Columbian mammoth, we know from its skeleton that it was very similar to these tundra beasts. It was a larger animal, however, some estimates suggesting that the biggest males (and only a small sample has been examined) weighed an astounding thirteen tonnes, making it a contender for the largest elephant ever. In comparison, the biggest African elephant recorded was an exceptional ten tonnes, though a six-tonne bull is considered a large one. At over 3.4 metres high at the shoulder, with long incurved tusks, the Columbian mammoth would have been the most majestic sight on the prairie.[1]

Like modern elephants these creatures lived in family groups dominated by older females. We know this because of some ingenious studies of tusks undertaken by Dan Fisher of the University of Michigan. Fisher has demonstrated that, at around fifteen to twenty-five years of age, maturing male Columbian mammoths underwent a period of stress that stunted their growth (including that of their tusks). Apparently this occurred when they were expelled from the female-dominated family herd (just as young male elephants still are today) and forced to forage alone.

In the more forested environments of the eastern part of the continent we might have seen the American mastodon. These primitive relatives of elephants appear to have been more solitary than the mammoths. Their teeth were made up of five pairs of 'teats', giving these impressive creatures their name (mastodon, meaning 'breast-tooth' in

Greek). Fossilised stomach contents tell us that these unusual teeth were used to grind up the leaves and twigs of conifers (including spruce), leaves of deciduous trees, swamp vegetation and grass. The mastodons were ponderous beasts, nearly three metres high at the shoulder and about 4.5 metres long. They were much more heavily built than the mammoths and may have been slower moving.

While these two kinds of elephant-like animals might have first appeared to dominate the landscape, an amazing variety of other species would also have been evident. Long-horned bison were among the most abundant. They were bigger than the American bison of today and their horns were enormous and straight. Fossil horn cores (the bony inner part of the horn) can span over two metres from tip to tip, indicating that with their sheaths they could have extended to about two and a half metres. Mummies found in Alaska reveal that these beasts were a uniform brown with blackish tips in some areas. Detailed studies of their anatomy indicate that they associated in small herds, the males probably defending a territory much like present-day African wildebeest. In terms of their ecological niche they were the American counterpart of the Cape buffalo so abundant in African game parks today.[2]

Among the smaller ungulates we might have seen a creature resembling a dromedary but with legs about 20 per cent longer. This animal, appropriately named yesterday's camel, was one of the last survivors of a true North American group, the Tylopoda (including camels, llamas, oreodonts and other extinct herbivores) that first evolved on the continent 50 million years earlier. Herds of another species of tylopod, the large-headed llama, might also have been seen. It was larger than the living llama though similar in shape and ecology. Another venerable American group was well represented, for at least three species of horse inhabited the continent. Perhaps they divided up the ecological niche much as horses, asses and zebras do, but there is nowhere on

Earth today where one can see these species living naturally side by side.

The most common large mammal was a long-legged pig-like creature known as the flat-headed peccary that was about the size of a European wild boar. In the 1970s a related species was found living in the remote Chaco region of South America. Known as the Chacoan peccary, it evolved from ancestors resembling the flat-headed peccary that migrated to South America between one and two million years ago. This living relic is now endangered, but a study of its ecology may tell us much about its extinct North American relative.

Along with these abundant and obvious species were a host of rarer large mammals, some of which may have looked strangely familiar. In the north the stag-moose (*Cervalces*), an animal about the size of a moose but with more deer-like antlers, filled the ecological role occupied by the moose today. Extinct mountain deer (*Navahoceras*), the size of a wapiti, would have resembled living deer with a twist, while the dog-sized four-horned pronghorn (*Breameryx*) could have been seen alongside its larger relative (*Stockoceros*) and the still-living American pronghorn. Forest-dwelling musk ox (*Bootherium*), larger than the surviving species, would also have fallen into this category of 'similar yet different'.[3]

Much more bizarre would have been the giant sloths, the largest of which were the megatheres (such as *Eremotherium*). I find it hard to conjure these creatures in my imagination, for they defy most efforts at making them appear believable. Imagine a creature six metres long from nose to tail-tip, weighing three tonnes and covered with coarse shaggy fur, in whose skin countless thousands of tiny pebble-like bones are embedded. The forefeet bore enormous claws, the longest over thirty centimetres in length, while the flat hindfeet were like great plates, each about a metre long and ornamented with gargantuan claws. To top it all, these creatures could raise themselves upright, reaching high into the trees to feed. The largest, paradoxically, were inhabitants not of dense forest but of open country, where they fed primarily on coarse bushes and grass.

While all ground sloths were weird, not all were so large. The North American nothrothere weighed a mere 180 kilograms—no more than a lion. One spectacularly preserved individual, pickled in dry bat guano that accumulated in a cave in the south-west, is on display in the Peabody Museum at Yale. Its skeleton still has tendons and patches of skin and fur adhering to it, although hungry rodents that fell into the cave ate away much of the body over the years. Caves full of sloth dung have also been discovered in the American west, the examination of which confirms it was a browser, feeding on the roots, stems, flowers and fruits of desert plants including the formidable creosote. Other sloths (and there were at least five kinds) were intermediate in size, each likely to have specialised in a particular vegetation type or habitat. Sloth relatives such as the tank-like Mexican glyptodonts (I think of them as honorary turtles despite their mammalian ancestry), armadillo-like pampatheres, along with true, giant armadillos were part of this 'strange but true' fauna.

This odd-looking fauna was tasty fodder for a wide range of carnivores. Among the most common was the dire wolf, larger and more heavily built than the living grey wolf. The dire wolf was probably a pack hunter and scavenger, capable of killing the smaller sloths, llamas, deer and bison.

Dwarfing these impressive carnivores, however, were the short-faced bears. Unlike grizzlies and black bears, short-faced bears were pursuit predators that ran down their prey in open country, much as wolves do today. They are a North American production and, at around 700 kilograms in weight, one species of short-faced bear was the largest meat-eating mammal ever to have trod the Earth. The only surviving member of this lineage is the spectacled bear of South America, but it is a herbivore that split off early in the group's evolution. The spectacled bears evolved in North America and persisted there until about 13,000 years ago when a variant very similar to but 50 per cent larger than the living species inhabited Florida. The living species may in fact be a direct, dwarfed descendant of this North American giant.

The great bears and wolves did not have the carnivore niche to themselves, however, for members of the cat family were also numerous. The lion (evidently the same as the living species) was then common in Alaska north of the ice, while to the south lived an even larger kind, the American lion, which is classified as a distinct species (*Panthera atrox*) by some researchers. These were the largest lions ever, reaching half a tonne in weight (a large male African lion by comparison weighs just 250 kilograms). Despite their size they may have possessed small manes, as do Asiatic lions today. While their skeletal remains are quite common, more poignant reminders of these impressive creatures have been found. Deep in Cat Track Cave in Missouri, plate-sized footprints have been found in soft mud, exactly as they were left over 13,000 years ago.

Other cats of the ice age include the sabretooths and scimitar cats. They used their razored canines either to bleed their prey to death or to disembowel it. The sabretooth was a powerful, lion-sized cat that may have been able to kill adult sloths and half-grown mammoths. Its short legs indicate that it was an ambush predator, incapable of running down its prey. The scimitar cat was longer-limbed, and its brain had enlarged optic lobes, suggesting that it had acute eyesight. A cave full of the chewed remains of infant mastodons along with complete skeletons of young scimitar cats suggest that this species may have specialised in running down and killing the young of the largest mammals, then dragging the bodies to a lair to feed their young.

By the end of the last glacial maximum, sabretooths and scimitar cats were relicts in the New World. Both had evolved from Eurasian or

African ancestors and their stay-at-home relatives became extinct hundreds of thousands of years before the American species vanished. Some researchers have suggested that competition with hunting and scavenging human ancestors was responsible for their early extinction in the Old World.

The New World cheetah doubtless stalked the magnificent American pronghorn across the plains of ice-age America, while in the cold northern forests jaguars hunted their prey—perhaps deer and porcupines. Food and humans, not cold, it seems, have restricted these versatile predators to their current southerly distribution. Other cat species that followed those of the jaguars and cheetahs in size were the cougar (*Felis concolor*, 36–100 kg), the ocelot (*Felis pardalis*, 11–15 kg), the Canada lynx (*Lynx canadensis*, 5–17 kg), the bobcat (*Lynx rufus*, 4–15 kg), the jaguarundi (*Felis yagouaroundi*, 4–9 kg), the river cat (*Felis amnicola*, 3–7 kg) and the margay (*Felis weidii*, 2–3 kg). This spectacular array of felids 13,000 years ago exceeds anything seen elsewhere in the world and speaks eloquently of the biomass of herbivores that, during a moment in geological time just past, roamed the continent.[4]

North America's flesh-eating birds are an even sadder reminder of a once great army of meat-eaters than its surviving cats. The very largest were a New World group called the teratorns. These birds evolved in South America, but various species existed in North America until around 13,000 years ago, the largest of which, *Teratornis incredibilis*, had a wingspan of nearly five metres. This is a stupendous size by the yardstick of any living bird, but *Teratornis incredibilis* was just a baby in the family. One South American species had a wingspan of up to twelve metres— the size of a small Cessna! This extraordinary bird, known as *Argentavis magnificens*, weighed around 80 kilograms—as much as an adult human. Such large birds are a puzzle to scientists and many aspects of their lifestyle remain obscure. There is still considerable debate as to whether any were active predators but the largest, surely, must have played a scavenging role, being too big to swoop to catch its prey.[5]

New World vultures (including the condor, turkey buzzard and black buzzard) are a rather unique New World group of large avian flesh-eaters. Recent biochemical studies have shown that they are not related to the Old World vultures at all, but to storks. Although this group probably originated in South America, one subgroup, the condors, appears to be truly North American in origin. Condors are the largest of the living New World vultures and the Andean condor is the largest flying bird on Earth. *Breagyps* is an extinct, long-beaked scavenger similar in size to the Andean species that wheeled in the North American skies over mammoth and mastodon just 13,000 years ago. As its name suggests, its remains are abundant in the tar pits at La Brea, Los Angeles.[6]

The California condor (*Gymnogyps californianus*) is the sole North American survivor of this avian megafauna. Before the extinction of the large mammals on which it fed it was far more widespread than in historic times, living as far afield as New York. It survived the extermination of its food source at least in part because the marine megafauna (seals, whales and dolphins) so plentiful off the Californian coast did not share the same fate as their land-based brethren. Although condors often wandered far from the west coast, it seems likely that for 13,000 years they survived by scavenging from the beached remains of these ocean giants. Its unique survival makes the California condor a gift from the ice age to the modern world—an echo of this distinct and now vanished North American splendour.

A decade ago the California condor was almost lost to the world. In 1986 just three free-living individuals survived, and all were male. Two years earlier there had been fifteen wild birds, including five pairs. The condors suffered from a number of problems: DDT poisoned their eggs, their eggs were stolen, they ate baits put out for coyotes, they ate lead pellets from carcasses, they were shot. On top of all this, ever more efficient industrial agriculture left little food for them, for any farm animals that died were quickly removed from the fields, and sometimes ground up and fed to their living brethren. This strange form of pastoral cannibalism was soon to bring mad cow disease to the world and, through removing dead animals before the great birds could find them, looked set to deprive it of condors as well.

By the early 1980s it was clear that without heroic action the California condor would be lost. A panel of experts recommended a hands-on emergency program of captive breeding as the only hope. They were, however, almost immediately shouted down by some formidable opponents. Carl Koford, then the world expert on the species, claimed that 'handling, marking and caging greatly diminish the recreational value of wild condors'. The nature writer Kenneth Brower also took the view that the condor should be allowed to exit gracefully. 'Perhaps feeding on ground squirrels, for a bird that once fed on mastodons,' he wrote, 'is too steep a fall from glory. If it is time for the condor to follow the *Teratornis*, it should go unburdened by radio transmitters.'

Despite the protests the recovery program went ahead and in April 1987 the last wild condor, a seven-year-old male (unpoetically known as AC-9) was taken into captivity. For years the success of the project was uncertain, but eventually the strategy of breeding the species in confinement then releasing the young paid off, and now wild condors, better protected than before, can again be seen in Californian skies. And, due to an adventurous translocation, after 13,000 years they once again soar over the Grand Canyon.[7]

When condors were at their peak over 13,000 years ago, extinct eagles, hawks, Old World vultures and teratorns shared the skies with them. The strangest in appearance was perhaps the long-legged 'walking eagle' (*Wetmoregyps*), an American version of Africa's secretary bird. The extinct Floridan hawk-eagle *Spizaetus grinnelli*, a near relative of the ornate hawk-eagle (*Spizaetus ornatus*) of South America, scanned the grasslands for small creatures. The extinct errant eagle (*Neogyps errans*) hunted considerably larger prey. A small Old World vulture known as *Neophrontops* (a group not found in the Americas today) was similar to the Egyptian vulture *Neophron*. So rich was this assemblage of scavengers that it would have made Africa's vulture-laden plains look impoverished in comparison.

The vanishing of so many species left the large avian scavenging niche all but empty in North America. These extinctions were far more profound than those afflicting the mammalian carnivores as carrion-feeders like the La Brea stork (a relative of Europe's white stork) became extinct in North America, and golden eagle and bald eagle numbers declined dramatically. Even small scavenging raptors—such as a relative of the yellow-headed caracara, which today hunts over the grasslands and savanna of South and Central America—became extinct over much of its range at this time.[8]

To complete the list of bird species vanished from what is now the United States we must add the ocellated turkey (which survives in Yucatan), and the so-called 'mastodon birds' *Pandanaris* and *Pyelorhamphus*. These extinct, starling-sized perching birds resembled cowbirds in their habits and were probably dependent on a symbiotic relationship with elephants or other large creatures, much as oxpeckers are today. The demise of the megafauna spelled their end also.

Before leaving North America's vanished airborne legions, one last species must be mentioned, for just 13,000 years ago the continent was home to giant vampire bats. These creatures, which weighed perhaps twice as much as the surviving South American species, lived alongside normal-sized vampire bats and all doubtless fed off the blood of great mammals such as ground sloths. The giant vampires vanished when their last blood bank spilled onto the North American prairie (likely at the point of a stone-tipped spear), while their smaller relatives found blood and succour to the south.

We should also glance briefly at the oft-forgotten reptilian element in North America's extinct megafauna. Six species of giant land tortoise related to the Galapagos giants once roamed the continent, the carapaces of the largest reaching 1.2 metres in length. Even during the height of the last cold phase, various species had broad distributions across the more southerly portions of the continent.[9]

It requires great effort for the contemporary mind to envisage a complete, working megafauna. Even museum exhibitions such as those of the incomparable Page Museum at the La Brea tar pits in Los Angeles can only give us a small indication of what it was like. To really understand one has to walk with the megafauna, to smell and hear it, to feel its threat, and then to appreciate it.

In 1993 I was fortunate to journey, on foot, through the thorn-scrub thickets of Wankie National Park in Zimbabwe. Following the trail of great mammals through the scrub, I measured my own puny footprint against the crinkled plate-like spoor of a male elephant. It was so crisp in the soft dusty soil that the near-metre-wide impression must have been just minutes old. And there were great piles of dung, each pellet the size of a small soccer ball, piled high into a modest tenement for the delectation of dung beetles. The beetles, each as large as my thumb, worked feverishly on the pile, trying to reduce it to billiard-ball-sized spheres for burial to feed their larvae.

Buffalo scrapes, the pervasive odour of big game, the hoarse grunting of a nearby lioness, half an antelope killed just hours before by a leopard, the blood still fresh on the grass—all this reminded me that here I was a tiny and insignificant being, a potential prey item in a world of giants. Were it not for our solitary rifle and all the technology it stood for, our party could never have entered that world and returned. To think of an entire planet like this—before humans—is an awesome thing.

As will become clear later, this lost megafauna is of supreme importance to contemporary conservation biology. We must learn how such a vast number of large mammals interacted with the vegetation that supported them, for the way we interpret this dictates our response to such vital issues as feral animals, species reintroduction, culling and fire. It may even revolutionise our rangelands management.

Many researchers have looked to Africa to provide a model of a world with its megafauna intact, but we have learned that even its splendid animals do not represent a complete assemblage of large mammals. Neil Owen-Smith, the eminent South African biologist, and J. E. Danckwerts wrote: 'Today there is no region in Africa where both a grazing mega-herbivore and a browsing megaherbivore coexist at population levels limited by vegetation capacity.' In other words, the megaherbivores of Africa have lost parts of their diversity and distribution. Key species in the environment such as elephant and white rhino no longer overlap in vast parts of the country, and where they do some other vital species is missing, or one or the other is not present at the maximum level its environment could sustain. Africa nonetheless gives us a framework.[10]

In North America, as in Africa, the regional abundance of large mammals is closely related to rainfall. The total combined weight of mammals on the richest African savannas can exceed 10,000 kilograms per square kilometre, and such places typically receive more than 800 millimetres of rain per year. Much of eastern and north-western North America falls into or near this rainfall zone. Owen-Smith also noted that 'Megaherbivores (species exceeding 1,000 kilograms in weight) typically form 50% or more of the total weight of large herbivores in savanna communities'. In Africa the largest browsers and grazers comprise the versatile African elephant, the white and black rhino and the hippopotamus. African buffalo (like the bison) are the most abundant grazers, forming up to 35 per cent of the total weight of large herbivores. Although trees typically constitute half of the primary production on savannas in Africa, browsers (excluding elephants) form only 5 to

20 per cent of the biomass. By analogy the Columbian mammoth, mastodon and bison were probably common, while llamas and browsing sloths are likely to have been comparatively rare.[11]

There was of course much regional variation across North America during the time the megafauna existed. The limited megafauna present in the farthest north, on the steppe that developed to the north of the ice in Alaska, is of special interest for two reasons: because frozen carcasses have revealed many details of the biology of the large mammals, and because it gave rise to the megafauna surviving in North America today. Alaska's megafauna, unfortunately, was not distinctively American but an extension of that of northern Asia and Europe. This is because, during very cold periods, we can think of Alaska as forming part of Eurasia rather than North America, for ice cuts it off from the rest of its continent, while the broad Beringian plain opens contacts with Siberia.

In Alaska the woolly mammoth replaced the Columbian mammoth, horses (very similar to Przewalski's horse) were common, as were long-horned bison (related to the Eurasian bison), while caribou, stag-moose and musk ox were present together with rarer or smaller species. Among the carnivores were lion and wolf. With the exception of the mammoth, all of these species, or their very close relatives, have survived elsewhere.

These creatures lived in one of the most extreme environments occupied by large land mammals, and as a result some have (or had) a most peculiar biology. They are every bit as adapted to the seasonal change as the tundra plants, for they are physically incapable of growing during winter. Feed a musk ox all winter long on the most nutritious herbage you can find and still it will not grow. Furthermore, its horns and bones will bear the marks of stopped growth just as the rings of a tree do. Like all northern animals it must eat enough during the summer to survive right through the long winter night on the body fat it accumulates.[12]

Thus was North America before a human footprint ever marked it. In this third act comprising 30 million years of continental history we have watched the continent's crust stretch and buckle, seen its trees migrate the length of the land and its fauna change markedly. If I had to choose one defining feature of this diverse age, however, it would be the supreme importance of climate in dictating North American life, for it was during this period that the great American climatic trumpet first sounded. From that moment on, the relationship between temperature and biodiversity in North America was unequivocal. As the celestial thermostat is turned down, biodiversity in North America dwindles and immigration slows; as it is turned back up, biodiversity flourishes. This relationship is due in part to the continent's profound ability to amplify global climate change.

The second striking feature of the age has been the influence of Asia and its creatures, particularly mammals, as they swept in across Beringia, making North America a land of immigrants, an eternal frontier. One might expect that successive waves of immigration would have caused the immediate extinction of some of the older lineages, though this is rarely the case. Instead, the new invaders and the longer established fauna often co-existed for a considerable period, only for both to suffer dramatic extinctions as the global temperature dropped.

Arching over these phenomena is the observation that, as opposed to the situation on most continents, the deep past seems to leave a limited legacy in North America. Change in the form of climatic shifts and massive immigration has reshaped the continent's ecology continuously. It is as if the continent knows no rest, no equilibrium, no consistent state to which we can point and say, 'Ah, see, that is North America.'

We have watched the evolution of those truly North American creatures, the dogs, camels and horses. We have seen the results of contact between the two halves of the New World, a North American takeover of the south, and the more selective South American march northwards.

We have seen an ice age begin and a new fauna take shape—including mammoths, mastodons, great sloths and teratorns.

In the foregoing acts we have sampled only the very coarsest segments in the vast expanse of time. We have dealt in millions of years rather than centuries, and this has affected the way the story of the continent has opened to us. It has prevented us from asking questions about short-lived phenomena, for example, about the impact the very first elephants to arrive in America had on the vegetation, or the effects newly arrived predators had on their prey.

Next is a new realm of time; one where questions about critical events *can* be addressed. This new realm consists of the 13,200 or so years since the first Indian sighted his (or her) great new discovery—a whole New World.

Act 4

•

In Which America Is Discovered

THIRTEEN

A NEW WORLD

One day perhaps we will be able to state with certainty the decade, or even the year, that ushered in the peopling of North America. At present, however, the possibility that we will know this prehistoric equivalent of Columbus Day seems remote. The problems are twofold: the nature of radiocarbon dating, and disagreements among archaeologists on how to interpret the dates and other data.

One fundamental problem is that radiocarbon dates do not give us the real age of the object being dated. This goes back to the technique's invention in the 1940s by the University of Chicago's Professor Willard Libby. He had worked with the Manhattan Project (whose products would so disturb the atmosphere that dating samples more recent than Hiroshima is impossible). His idea for radiocarbon dating was considered so crackpot at the time that he was forced to work in secret, publishing his first dates in 1949. His methods do indeed appear eccentric, for he once descended into the Baltimore City sewerage system to obtain a methane sample—literally a fart in a bottle.

Libby found that Carbon-14 was constantly created in the upper atmosphere and that it decayed at a known and steady rate. Any living thing incorporates Carbon-14 into its body, but as soon as it dies its intake of Carbon-14 ceases and whatever has accumulated begins to decay. Libby realised that by measuring the ratio of Carbon-14 to Carbon-13 in the remains of a living thing he could estimate how much time had elapsed since it died.

Despite the fact that Libby was granted the 1960 Nobel Prize for his work, his findings were flawed: he miscalculated the rate of decay of Carbon-14 (the decay constant) by around 3 per cent and, contrary to his belief, Carbon-14 is not created at a steady rate in the upper atmosphere. This miscalculation was not detected until the late 1960s, and by that time many thousands of radiocarbon dates had been published. The difficulty of a reader knowing which dates were calculated using the incorrect decay constant, and which were calculated using a new one seemed so overwhelming that the scientists decided it was better to stick with the original, incorrect decay constant. Thus, every raw radiocarbon date you read today is given as too young by around 3 per cent.

This, however, is a simple problem compared with the creation of Carbon-14 in the upper atmosphere. It turns out that Carbon-14 is formed from nitrogen through the bombardment of the atmosphere by cosmic rays, and that the Earth's magnetic field regulates how much cosmic radiation reaches the atmosphere. We now know that the strength and polarity of the Earth's magnetic field fluctuates through time—sometimes quite rapidly. Thus more Carbon-14 is created in some periods than others.

In comparing the results of various dating techniques, scientists have now developed a correctional factor that can be applied to a radiocarbon date (plus its 3 per cent) to give an estimate of the real age of the object being dated. For radiocarbon dates of around 11,000 years one needs to add around 2000 years to obtain the real age. For dates of a few thousand years—as well, surprisingly, for dates of greater than 30,000 years—one needs to add fewer years as the radiocarbon and real dates converge. And a radiocarbon date of around 40,000 years (the upper limit of the technique) actually gives close to the real age of the object!

An additional problem arises concerning the degree of precision with which a date can be given, and the period in which we are most interested here (around 13,000 years ago) is one during which poor precision prevails. Unless otherwise specified, dates given here are in real years (that is, where the various correctional factors have been applied to raw radiocarbon dates), but it is as well to remember that that 'around 13,000 years ago' implies less precision than, say, 'around 6000 years ago'.

Archaeologists are split into two camps on the question of when people first entered the Americas. One camp (which I fear represents the minority at present), thinks that the Clovis people, whose artefacts date to around 13,200 years ago (that is 11,200 radiocarbon years), were the first people to enter North America. The Clovis culture, incidentally, is named after a town in eastern New Mexico near which Clovis artefacts were discovered in the 1930s. The other camp argues for an earlier human presence (some would say tens of thousands of years earlier) by a mysterious group referred to as pre-Clovis people.

Despite decades of intensive research, the likelihood of a pre-Clovis presence in North America remains contentious. Dr Tom Dillehay is one researcher who thinks that there were people in the Americas before Clovis. He is also dismayed at the lack of objectivity that sometimes infuses the debate, saying that it has degenerated into 'a sport in the study of the first Americans', which he characterises as a vicious game of hurling emotional and injurious opinions in public. It is also a sport that has confused many non-experts, for unsubstantiated claims about the antiquity or otherwise of a human presence at various sites have often been published as if they were fact.[1]

Many sites thought to predate Clovis in the Americas—such as Calico,

Louisville, Old Crow, Texas Street and others—have now conclusively been shown to postdate the Clovis culture. And there are a few others, known colloquially as 'the ones that won't go away', which have produced dates for human occupation in excess of 13,200 years, and which remain contested.

One of the most extraordinary of these is the Monte Verde site in southern Chile. There, bones of the South American elephant-like gomphothere (*Cuvieronius*) have been found together with pieces of animal skin (attached to poles in some cases), wooden and stone tools and even meat fragments. The skin is clearly from a large mammal— possibly *Cuvieronius* itself. If so, it is of exceptional importance to the science of palaeoecology, for it could tell us much about the appearance of these now-vanished creatures. Such delicate things as a human footprint have even been discovered. The remarkable preservation at the Monte Verde site is due to its becoming waterlogged and sealed by a layer of peat. Eleven radiocarbon dates on wood and bone suggest that the site was occupied between about 14,800 and 14,400 years ago.

Let us assume for a moment that Monte Verde, as agreed by a blue-ribbon panel of archaeologists, indicates a pre-Clovis presence. This implies that humans had been in the New World for at least 1200 years before Clovis, leaving little trace of themselves except for a few other problematic sites such as Meadowcroft Shelter in Pennsylvania (supposedly occupied at least 19,000 years ago). Furthermore, these people had no detectable impact on the megafauna or on vegetation through fire.

Champions of a pre-Clovis presence hypothesise that evidence of these people is rare because they possessed an unsophisticated stone toolkit of pebble choppers, scrapers and knives and that the cool, dry conditions of the time kept human numbers low. We can test this hypothesis by looking elsewhere for people who live or have lived under similar conditions, and determine whether their archaeological record is as elusive as the supposed pre-Clovis cultures. The best match for this hypothetical pre-Clovis culture is found in Australia's Aborigines. For the past 50,000 years they have inhabited a dry continent at what is, by global standards, a very low population density (there were perhaps only 500,000 in a landmass the size of the contiguous forty-eight states), and they used a very basic, if multi-functional and elegant, stone toolkit.[2]

Using this yardstick, the arguments of the pre-Clovis proponents are unconvincing, for Australia's Aborigines have left abundant, continent-wide evidence of their presence between 20,000 and 40,000 years ago. The Australian example indicates that if people were present in North America more than 13,200 years ago, we should find plenty of evidence of them.

Other proposals to explain the existence of people at specific sites, such as Monte Verde, before 13,200 years ago include the suggestion that the Monte Verdeans sailed the entire length of the west coast of the Americas—from Alaska to Chile—before deciding to settle. To imagine people sailing such a vast distance, past coasts full of game animals and other resources but not stopping, their hearts resolutely set on southern Chile, strains credulity. After all, on the way they would have bypassed California, the region that supported the largest numbers of Indians north of Mexico, and all to settle in cold, inhospitable southern Chile!

That being said, I don't know what to make of the Monte Verde site. The second volume of the site report, written by Tom Dillehay, comprises nearly 1100 closely argued pages setting out dates, descriptions of artefacts and site stratigraphy. One can always come up with criticisms, but until a careful, independent analysis is undertaken, such criticisms are unhelpful. One thing that makes me cautious about accepting the results is that Dillehay also finds evidence that people lived near Monte Verde around 33,000 years ago. If true, this magnifies all of the problems alluded to above.

It is clear that either Tom Dillehay and the other supporters of a pre-Clovis presence in the Americas are wrong, or the majority of archaeologists worldwide are misled about global patterns in prehistory.

Although lacking a convincing explanation for the site, I am one of the Monte Verde (and thus pre-Clovis) sceptics, and from here on will write as if reports of a pre-Clovis occupation of the Americas result from dating or other interpretive error.

If the Clovis people really were the first American pioneers, the Americas were the last of the habitable continents to be occupied by humans, being settled around 40,000 years later than even Australia. Given the existence of the Bering land bridge it seems extraordinary that the New World remained hidden for so long.

Archaeologists suggest three potential barriers may have shielded the Americas from humanity. The first is located in Eurasia—the extreme climatic conditions that exist as one approaches the Arctic Circle. The second barrier is the Bering Strait. The third lies within North America— the great continental ice sheet that divided Alaska from the rest of ice-free North America until about 13,500 years ago.

While logic based on modern geography might suggest that the Bering Strait was the fundamental barrier, this is almost certainly not the case, for the world has changed dramatically over the last 14,000 years. Then, there was no Bering Strait, but instead a vast ice-free plain 1600 kilometres wide from north to south connecting Eurasia and Alaska. In fact, study of the archaeological record reveals that the extreme climate of northern Eurasia was probably the critical barrier. Before 14,000 years ago conditions were even more severe than today and no woody plants at all grew at high latitudes. People only gradually encroached on this formidable realm: 30,000 years ago there were no humans living north of 54 degrees, while even 15,000 years ago few if any people ventured north of 60 degrees anywhere in the northern hemisphere. Then, the tundra was a vast, if not entirely safe game reserve from which humans were for the most part excluded. There is, however, evidence that people living in forest bordering the tundra would venture north in summer to harvest mammoth and other megafauna before winter forced them to retreat south. Despite these incursions, extensive areas of tundra were too distant from the trees ever to have borne a human footfall.

One innovation essential for any people bent on conquering the far north is warm clothing. Evidence of such perishable material is rare in the fossil record, but a remarkable find made at Sungir, near Moscow, reveals that clothing of a very advanced kind was manufactured there at least 20,000 years ago. The skeletons of a man and two boys excavated from a grave were covered with decorative rows of ivory beads that researchers realised were once part of a costume. Painstaking reconstructions reveal that their clothes consisted of caps, shirts, jackets, trousers and moccasins. With suitable clothing being made at such an early date, it was clearly not lack of covering but some other factor that kept people out of the tundra.

Fifteen thousand years ago conditions were even drier and colder than they are today, and there were no trees or shrubs in the north. At that time the sophisticated Eskimo cultures that can exist virtually without timber were still far in the future. It may be that the availability of wood, for tools and fire, made the critical difference to the stone-age humans living at the northern edge of the inhabited world. If so they were in luck, for the Earth soon began to warm rapidly and trees migrated north in response.[3]

The people of north-eastern Eurasia at this time had developed two effective means of killing large mammals—the spear thrower and drive hunting. The last mammoth, woolly rhino and Irish elk in the northern mainland of Eurasia date to about 14,400 years ago. The coincidence in timing between the human northern expansion, improving technology and Eurasian megafaunal extinctions is close, suggesting that the humans on their way to Beringia may have already had an ugly track record as exterminators of big game. Some Eurasian megafauna managed to survive on a few human-free island refuges such as Ireland, where the Irish elk persisted until people arrived about 12,000 years ago;

the islands of the Mediterranean, which supported pygmy hippos, elephants and other curious beasts until settled by humans over the last 10,000 years and Wrangel off Siberia, where mammoth survived until about 4000 years ago.[4]

Sometime between 15,500 and 12,500 years ago people settled on the shore of the Arctic Ocean at Berelekh in north-eastern Siberia. There they hunted mammoth; more than 7000 bones from these great beasts have been found near their campsite. There are about 200 bones in each mammoth, so those bones represent a lot of mammoth steaks—about 250 tonnes of prime cuts to be precise.

Berelekh must have been directly on or just adjacent to the migration route that took people to the New World. The material culture of the first Americans shows marked similarities with that of the people of eastern Europe and central Asia, reflecting this common ancestry. We can imagine the inhabitants of Berelekh pushing ever further east in their pursuit of dwindling game, until they finally crossed the still-wide plains of Beringia some 13,800 years ago. Upon arriving in what is now Alaska they would have found more of the same—mammoth, bison, horse and reindeer—a fauna they had been eating their way through ever since their ancestors broke out onto the tundra a millennium earlier.[5]

Alaska, however, acts as a revolving door. When Beringia is above the waves it opens east, but then its portal south is closed by a great ice cap. As the ice melts it floods Beringia, closing the western entrance as the ice blocking the southern one retreats. At first the gateway south would have consisted of a narrow corridor between the retreating Laurentide and Cordilleran ice caps. It was an inhospitable, frigid and boggy corridor largely devoid of life. People may have shunned it for centuries, but finally someone decided to explore it. After threading their way past freezing mires, under louring walls of ice and through dense fogs for a hundred miles or more, those early explorers emerged somewhere near present-day Edmonton. To the south lay an enormous, megafauna-dotted plain. A whole New World, stretching almost from pole to pole, lay before them.

What did these first Americans look like? Christopher Columbus might have been confused when he called the people he met 'Indians', but the name was felicitous. Perhaps he noticed a resemblance to the Asian people of the East Indies in their coarse, straight black hair, light-brown skin, brown eyes and high cheekbones. Had he examined them more closely he would have found even more shared characteristics: their incisor teeth are hollowed out behind (in anthropological jargon they are shovel-shaped), and their infants often develop a purplish colouration on the lower back (the so-called Mongoloid spot). These traits, coupled with close genetic similarities, indicate that America's Indians are an Asiatic people.

Despite the many physical resemblances, there are differences between American Indians and most contemporary Asians. Few Indians, for example, possess the fold of skin over the corner of the eye that is such a striking feature of most modern people from north Asia. A further difference is found in the fact that many American Indians have prominent noses and narrow faces, while most contemporary north Asians have small noses and broad faces. The skull shape of American Indians is also more variable (particularly if one includes the skulls of older prehistoric Indians) than are those of north Asians. Physical anthropologists suggest that these differences indicate that the American Indians left north-east Asia before all the characteristics of the modern Asian population had become established.

As these people entered their New World their toolkit underwent an astonishing transformation. Almost instantly they began manufacturing an entirely new kind of instrument, the Clovis point. These magnificent fluted stone spearheads are of distinctive design, ranging from around four to twenty-three centimetres in length. Highly functional, they are also works of art. Looking at them one senses a vanished pride, a sense of great achievement expressed in the strict functional beauty that delivered so much into the hands of their makers.

All other surviving Clovis culture artefacts, such as end-scrapers, shaft-straighteners and cutting tools, were brought with the invaders from Eurasia and were not significantly modified after their arrival. It is the Clovis point alone that has no antecedents in the toolkits of the Old World. They were the first American innovation, and without it I believe that the bounty of the first American frontier could not have been harvested. In that sense, they were the railways of the late Pleistocene.

Even though Clovis points indicate that their makers were big-game hunters, some archaeologists doubt whether meat from large mammals was that important to them. Surely, they surmise, plants must have played a vital role in the Clovis diet, just as it did in the diet of most later Indians. Yet the absence of grinding tools at Clovis sites suggests that they did not eat many seeds. The archaeologist Stuart Fiedel agrees that perhaps the Clovis people ate nuts and berries but, he says, 'it is hard to avoid the conclusion that the [Clovis] paleo-indians were more dependent on hunted meat, and less reliant on gathered plants, than later Archaic populations'.[6]

One intriguing uncertainty about the Clovis people is whether dogs accompanied them on their journey to the New World. Dogs may have been domesticated by around 16,000 years ago in the Middle East, and may well have reached north-eastern Asia by 14,000 years ago, in time to cross Beringia with the ancestors of the Clovis people. Although evidence of earlier remains are sometimes published, the earliest well-dated domestic dog remains in the New World are 10,500-year-old skeletons found at the Koster site, in the Illinois River Valley. Dogs may have been brought to the Americas by a subsequent group of invaders or visitors, but these people would probably have had to travel to the New World by boat, or across sea ice, for by then Beringia had effectively submerged below the sea.

The Clovis people made their stone points for an extraordinarily brief period—300 years is what has been determined from uncalibrated radiocarbon dates. These lethal yet elegant sculptures became useless around 12,900 years ago, I believe because, as will soon become clear, the last of the very large creatures they were designed to kill had all fallen victim to their efficiency.[7]

Clovis points, and what we know of Clovis culture as a whole, show relatively little regional variation. From Guatemala to the Dakotas, and from the Pacific to the Atlantic coast, the method of manufacture of Clovis points and other artefacts was uniform. Unless we count our own time, with its ubiquitous Coca-Cola cans and baseball caps, such cultural homogeneity has never been seen since in North America. It suggests that the Clovis people occupied the entire continent within a century or two and that the impetus for manufacturing their spear points existed over the entire region. This and the brevity of the culture's existence strongly suggest that Clovis was a pioneer culture. The fact that functional tools comprise almost all that Clovis people left to posterity also speaks eloquently of a frontier existence.

What is really striking about the Clovis people, however, is not what they possessed, but what they apparently lacked. There is no evidence, for example, that they made the distinctive and rather sophisticated houses used by their ancestors in Eurasia, and many researchers suspect that their habitations were rudimentary. They probably sought shelter under simple, portable tents made of sticks and animal skins. This is a considerable step down from the abodes of their Eurasian ancestors, who built elaborate semi-subterranean huts and used mammoth bone to construct sturdy above-ground structures.

There is also little evidence of elaborate burials, which were common in Eurasia at the time. Clovis people sprinkled their dead with red ochre as did their relatives in Eurasia. Nonetheless, Clovis burials show little sign of the elaborate ritual evident in contemporary Eurasian gravesites. Interestingly, the use of ochre is lost in the Americas soon after Clovis vanishes.[8]

The most astonishing thing of all, however, is the fact that the Clovis people produced no figurative art, or at least none that has survived to the present. The only possible exceptions are two bone carvings, both from Mexico. One is the head of a coyote-like animal carved from a camel vertebra that was found by a farmer digging a drain in 1870. The other is the pelvis of a mastodon with some rude scratchings on it that may represent a tapir, mastodon or bison. Alternatively it may be a meaningless meander. In any case, large questions hang over the age of both of these carvings, and they may well be the work of a later people.

This absence of art seems odd indeed when contrasted with the 'high' cultures of Eurasia from which Clovis sprang. Consider, for example, the great and ancient art tradition that had developed in northern Eurasia 20,000 years before the Clovis people entered North America, of which the caves of Lascaux and Altimera, with their vivid depictions of a vanished megafauna, are the most famous. Elsewhere in Eurasia hundreds of depictions of the mammoth, other megafauna and even humans have been found in Clovis-age and older cave art, stone and ivory carvings and even clay figurines. And in considering the artistry of the ancient Eurasians we must remember their superb craftsmanship in fashioning jewellery. Some of these art traditions were widespread, and even as far east as Berelekh people were carving elegant representations of mammoths on ivory.

Thus Eurasia has an artistic tradition that both predates and postdates Clovis, yet not a single representation—not one object that might be called art—has been found that can be convincingly attributed to a Clovis artisan. It is possible of course that Clovis artisans wrought wonders in perishable materials such as wood, leather, feathers and fur, but it strains credulity to think that all of their artistic efforts were channelled into works that would not last. I am not suggesting that the Clovis people lacked artistic sensibility, for the beauty of their stone points and cross-hatched ivory points is beyond dispute. It's just that their sense of beauty appears to have been entirely functional. It is as if they had no desire to represent their world in art; as if for them, the function of art was to help conquer their environment. Theirs was, I feel, an aesthetic that would find beauty in a well-oiled engine.

If we cast about for other cultures possessed of similar characteristics, does any more readily suggest itself than that of the American pioneer? Both rapidly peopled North America with a uniform culture, both were preoccupied with function, and both will be remembered for their utilitarian products. For Clovis it was a new kind of hunting point, a spear tip that allowed them to exploit the new and enormous frontier resources of the North American megafauna. For the later pioneers it was their firearms and wagons. Even the duration of the pioneer phase of both cultures is similar: 300 years, between about 13,200 and 12,900 years ago for Clovis, and about the same between 1607 and 1890, for the North American historic frontier.

Some pioneers of the historical period left a written record, telling us why they acted as they did. In 1743 Benjamin Franklin observed that his colleagues did not write books because they were too busy with other things. One wonders whether a Clovis hunter would have said the same thing if asked about ivory carving. It seems that for these two very different pioneering cultures, it was (to paraphrase the historian Frederick Jackson Turner) the wilderness that mastered the colonist.[9]

FOURTEEN

THE BLACK HOLE THEORY
OF EXTINCTION

Since the Alaskan revolving door last opened about 13,500 years ago admitting humanity to the Americas, much has changed. As long ago as 1839 Charles Darwin wrote, 'It is impossible to reflect on the state of the American continent without astonishment. Formerly it must have swarmed with great monsters; now we find mere pygmies compared with the antecedent, allied races.' Darwin's co-discoverer of the theory of evolution via natural selection, Alfred Russel Wallace, added to the insight by recognising that the phenomenon was a near-global one, stating, 'we live in a zoologically impoverished world, from which all the hugest, and fiercest, and strangest forms have recently disappeared'.[1]

Wallace called the extinction of the giants a 'marvellous fact', and over the century since the fact was recognised it has proved to be as difficult to explain as it is marvellous. A host of causes have been mooted, from 'racial senility' to disease, cosmic rays and even God's desire to clear the way for His chosen people and their herds, but in recent years two separate explanations have come to dominate the debate. On one side are those who argue for a climatic cause, while on the other are those who hold human hunting to be responsible. I subscribe to the black hole theory of extinction, which suggests that at the end of the Pleistocene all of those huge mammoths, sloths, camels and mastodons disappeared into a black hole. Not any black hole, mind you, but that very discerning one lying between nose and chin on the Clovis physiognomy.

At present, however, the majority of US scientists seem to favour global climatic change as the cause. Many theories as to how this might have occurred have been put forward. One of the most vigorously defended has been advanced by Dr Russ Graham of the Denver Museum and Professor Ernest Lundelius of the University of Texas, Austin. They suggest that the environment was more 'patchy' 14,000 years ago than it is today. As it began to assume its modern, rather monotonous form, competition between species increased. The evolution of new plant communities that resulted as plants changed their distribution in response to the retreat of the ice would have added another burden, as would a drop in the quality of plant food along with a rise in toxic plant species. 'Given the complexity of these systems,' Graham and Lundelius say, 'it is possible that some herbivore species literally poisoned themselves into extinction.' Their main point is that all of this climate-induced change at the end of the glacial maximum meant the end of the North American megafauna.[2]

Until very recently, the precise time that the continent's fiercest and strangest creatures became extinct has been difficult to pinpoint. It has been known since the 1980s that the North American horses, a camel, the Shasta ground sloth, the sabretooth, a tapir, a giant sloth, a giant beaver and the flat-headed peccary survived until around 13,000 years ago. Studies now under way by Russ Graham, however, indicate that with three vital exceptions, almost all of the more common North American late Pleistocene megafauna perished about 13,200 years ago. The exceptions are the mighty mammoth, the mastodon and the short-faced bear, all of which held on longer, the first two until about 12,800 years ago and the bear for perhaps 1000 years after that.[3]

The cause of the extinctions has been exceptionally tricky to resolve in North America because the time of extinction overlaps both a critical one for climate change and the time that the Clovis people flourished. The key to deciphering the riddle of the North American extinctions, I believe, is to examine extinctions in a global context. Because the intensity and timing of megafaunal extinction varied from continent to continent, we can ask whether the pattern is consistent with global climate change or the spread of humanity, for these present very different patterns worldwide.

In the oceans, in Africa and in south-east Asia the extinctions were non-existent or mild. Researchers have argued that this was because Africa was where humanity first evolved, so the fauna had time to adapt to the human presence. Likewise, they suggest, humans or pre-human species such as *Homo erectus* have long been present in south-east Asia, while until recently the oceans were inaccessible to people. Eurasia experienced moderate extinction rates, with most occurring long ago (when *Homo erectus* arrived there) or around 14,400 years ago when the tundra-dwelling woolly mammoth, woolly rhino and giant elk vanished.

The Americas, Australia, Madagascar and many oceanic islands, on the other hand, suffered dramatic extinctions. North America lost 73 per cent of all genera weighing more than forty-four kilograms, but Australia lost every terrestrial vertebrate species larger than a human as well as many smaller mammals, reptiles and flightless birds, the latter down to about a kilogram in weight. In all, about sixty vertebrate species were lost, including bizarre marsupials that resembled giant sloths and oversized capybaras, carnivorous kangaroos, a five-metre-long monitor lizard that was probably top predator and a terrestrial horned tortoise approaching a Volkswagen Beetle in size.

Australia is a vital natural laboratory in which to test extinction theories. Humans arrived there about 53,000 years ago. The climate was at its most extreme (dry and cold) between 25,000 to 15,000 years ago, and the rate of climate change was greatest between 15,000 and 10,000 years ago. The question then is whether the Australian megafauna died out during the time of dramatic aridity, or 20,000 years earlier when humans arrived.[4]

Establishing just when the marsupial giants last trod Australia's outback has been a tortuous business, with many false leads. For decades it was believed that the megafauna survived until close to the time of the last glacial maximum, some 20,000 years ago, when temperatures were up to nine degrees Celsius cooler than at present in the interior and the continent was extremely arid. Conditions were so extreme that trees all but disappeared inland and 40 per cent of Australia was transformed into a vast active dune field. Even parts of Tasmania were desert.

In early 1999 a breakthrough in the dating was made. It resulted from study of the remains of a bird called *Genyornis*. Weighing around 80 to 100 kilograms, *Genyornis* was twice as heavy as the largest living emu and cassowaries and an inhabitant of Australia's inland plains and some coastal regions. Its legs were short and thick, indicating that it was a slow runner. Proponents of human-caused extinction suggest that these characteristics would have made *Genyornis* vulnerable to human hunting. What is important about *Genyornis*, however, is that thousands of its eggshell fragments litter the landscape, and bird eggshell has unique properties that make it ideal to date using a technique called amino acid racemisation.

The 1999 study, by Professor Gifford Miller of the University of Colorado at Boulder and his colleagues, dated 1200 eggshell fragments, 600 each from emu and *Genyornis*, collected at four very different regions in south-eastern Australia. Their study demonstrated that *Genyornis* died out abruptly, probably right across the continent, around 50,000 years ago.[5] This is very close to the accepted time of arrival of people in Australia. It is still unknown whether the extinction of *Genyornis* coincided with that of the rest of the Australian megafauna, or whether just this one species went extinct at the time of human arrival. At present, however, there is no unequivocal evidence for the survival of any Australian megafauna more recently than 40,000 years ago.

The tale told by the *Genyornis* eggshells represents a serious challenge to those who believe megafaunal extinction was caused by climate. The period about 50,000 years ago was one of modest climate change (just a little cooling), well before the dramatic climatic fluctuations of the glacial maximum. Those supporting the climatic argument must now posit a climatic event that could have devastated megafauna well before the glacial maximum in Australia as well as just after it in the Americas. They must also explain how climatic change could have extirpated the large emu-like *Genyornis* in Australia 50,000 years ago, but leave New Zealand's eleven species of moa living just across the Tasman Sea unaffected until 800 years ago when the Maori arrived.

A new school of thought has recently entered into the extinction debate. It postulates that a *combination* of human impact and climate change was responsible the extinction of the world's megafauna. The *Genyornis* data weakens that argument, for the following reason: 50,000 years ago Australia experienced mild cooling while 13,000 years ago the Americas experienced rapid warming. These disparate climatic conditions were both coincident with megafaunal extinction, yet are so different that it is hard to imagine a climatic regime outside these extremes under which humans could arrive and large mammals survive. Thus the influence of climate in this hybrid proposal seems very weak indeed.

One final and very important challenge to the climatic theory of extinction has recently emerged. In 1999 researchers interested in the fossil fauna of the Caribbean published rigorously tested radiocarbon dates that give a 'time of last occurrence' for the ground sloths that once inhabited Cuba. These dates indicate that the creatures survived until at least 6250 years ago—around the time people reached the island. What climatic phenomenon, the advocates of the climate model must ask, could drive ground sloths to extinction from Canada to Tierra del Fuego in an instant of time around 13,200 years ago, yet leave the ground sloths of Cuba untouched? Indeed it is becoming evident that no extinctions at all occurred on any of the Caribbean islands 13,200 years ago, but that the region lost all of its large land mammals (mostly sloths and large rodents) and many bird and bat species around 6000 years ago. Again, extinctions correlate not with climate change but with the arrival of people. If no climatically based answer is forthcoming, this new data surely represents the final nail in the coffin of the climatic theory of megafaunal extinction.[6]

In discussing the possible causes of megafaunal extinction it is important to recognise that nothing quite like it had ever occurred before in the history of North America. The asteroid-caused extinction of 65 million years ago differed in that it removed a far greater variety of species, from small mammals and trees to plankton and larger marine invertebrates. Likewise, none of the other extinctions over the past 65 million years has so specifically targeted the large land animals.

As we have seen, the lesser, non-asteroid-caused extinctions in North America during the last 65 million years have been linked with climate change. This might predispose one to favour climate as the cause of the ice-age extinctions also, but there are important differences between the events. Paramount is the fact that the extinctions of 13,200 years ago occurred during a period of rapid warming, while all of the earlier ones have been linked with dramatic cooling. Furthermore, the species lost in these earlier extinction events are just those one would expect to be vulnerable to a cooling climate, such as browsers (which are affected by forest loss) and arboreal species. This correlation is not clearly evident in the case of the ice age extinctions. Why, for example, would a warming climate be bad for giant tortoises, peccaries, mixotoxodons and ground sloths?

It seems to me that seeking common causes or patterns in all extinctions is an exercise in futility. Leo Tolstoy began *Anna Karenina* with the observation, 'All happy families resemble one another, but each unhappy family is unhappy in its own way.' I think that ecosystems that have been stable for a long period are, like Tolstoy's happy families, all stable in the same way. Their stability is the result of being held in shape by the same basic ecological forces, which results in the evolution of primary producers, herbivores and carnivores in predictable numbers relative to one another. Indeed, astonishing similarities between elements in isolated ecosystems can be caused by these forces: witness the likeness of the thylacine and the wolf. Extinction events, however, are much more like Tolstoy's unhappy families: each extinction happens in its own way. This is because misfortunes are innumerable in this fragile world: they can fall from the sky, strike with a chilling climate, or come from the hands of man, and each has its own unique effect. Whether an extinction-causing event occurs in winter or summer matters, as does the technology possessed by an exterminating culture.

Those who champion humans as the cause of megafaunal extinction argue that the unique nature of the event is due to the power of our species. They argue that, almost from the moment our kind arose, our potential for destruction was unequalled in Earth history.

I believe that analysis of global extinction points towards humans rather than climate change as the cause of the demise of North America's giants, and that compelling evidence implicates Clovis hunters as the killers. The strongest direct evidence—the smoking gun if you wish—concerns sites containing both Clovis points and mammoth remains.

These sites indicate to me that the Clovis frontier was a meat frontier, and that when the last mammoth steak was finished at the last Pleistocene megafaunal barbecue, so too were Clovis and the North American megafauna.

FIFTEEN

MASSACRING THE MAMMOTH, DISMEMBERING THE MASTODON

At the time the first Clovis hunters emerged south of the North American ice sheet, mammoths and mastodons were widespread. Indeed their populations were probably still expanding as the great ice sheets covering boreal North America hastened their retreat. Mammoths could be found from central Alaska to deep into Mexico, with an outlier population in the east below the Great Lakes. Where mammoth were absent, such as in denser forests, one could find the cumbersome mastodons and gomphotheres. The American mastodon populated the eastern forests of North America, while the gomphotheres known as *Cuvieronius* inhabited the Gulf coast and Central and South America.[1]

The very abundance and expansive distribution of mammoths, mastodons and other large creatures make many researchers reluctant to accept the idea that the first Indians eliminated them. How, they quite rightly ask, could people armed only with stone-tipped spears have hunted such vast numbers of these magnificent animals off the face of two continents? Today there are just eight kinds of land creatures left on Earth that weigh one tonne or more: two species of elephant, four of rhino, the hippo and the giraffe. Just how difficult is it for people to kill such beasts?

A 1960s report on elephant culling in Kenya gives us some idea of how these largest of land animals react to being hunted:

> Three elephants, members of a herd comprising about 30 bulls, cows, immature animals, and calves, were shot in a large forest clearing. A fresh breeze was blowing from the direction of the herd towards where I was standing with a party of game scouts and because of this the confused and furious beasts could not tell from whence the danger lay. We were standing in full view of the herd at no more than 60 yards distance, and yet the elephants appeared to be unaware of our presence. They milled around trumpeting and shrieking fearfully, throwing up showers of grass and stones in their fury, and then with incredible determination and complete disregard for their own safety attempted to lift their dead companions. I watched this performance for a little more than half an hour, throughout which time the elephants persisted in their attempts at moving their stricken comrades, and showed no other interest or fear.[2]

This does not sound like the kind of creature one would hunt with a stone-tipped spear. It's worth remembering, however, that elephant hunting is a rather specialised profession in the modern world, and most of us, with the exception of a few experts in Africa and Asia, would have no idea how to go about it. We see the venture from an odd perspective, imagining a titanic head-on struggle between man and beast. The truth is that even modern native elephant hunters rarely tackle such formidable creatures head on. Instead they use tricks and ruses, or drive herds into dead ends or swampy ground. The same appears to have been true in Clovis America.[3]

In 1984, while still a visiting student, I went to the Lehner ranch in Arizona in the company of palaeontologist Paul S. Martin. Martin, tall and with striking blue eyes, is best known for his application of the term 'blitzkrieg' to explain the rapid extinction of the American megafauna at the hands of Clovis hunters. Despite the direction his research has taken him in, he has enormous respect for North America's indigenous cultures and I suspect that he is uncomfortable with some of the criticism his theory has brought him.

That morning we met Ed Lehner, frail but still fully engaged with the world, who led us to the erosion gully where he had found the bones of mammoths protruding from the soil forty years before. It was a day I shall always remember; a clear sunny morning in a dusty arroyo, with Lehner hobbling along as nimbly as his octogenarian legs would carry him, excitedly reliving the great dig that had occurred there so many years before. Beside me came the magnificent Paul Martin, making do with two sticks for crutches, on legs crippled long ago with polio. Despite their wonky pins the men's enthusiasm for the tale of the dead mammoths was undaunted.

From Lehner I learned that the site had been excavated in the late 1950s by the meticulous Emil Haury of the Arizona State Museum, who uncovered the bones of nine immature mammoth, among which were scattered eight Clovis points and four butchering tools. Most moving was Lehner's recollection that it was here, in 1959, that the first radio-carbon dates on mammoth hunting in the Americas were obtained. Haury and his team, along with Lehner, had been the first people to know that mammoth had stalked the Americas as recently as 13,000 years ago.

Visiting the arroyo with these two great men gave me an intense sense of the tradition of the continuing debate of American prehistory, of the uncovering of the bones and evidence. Their description of the gradual realisation that the mammoth had been surprised while drinking on a spring-fed stream and the young 'cut out' from the herd before being slaughtered, was revelatory even thirty years on. For me it was a day when not-so-distant prehistory hove into view.

At least eleven other localities where mammoths were slaughtered have been found in North America. They reveal that Clovis hunters often assailed the great beasts where they were most vulnerable—in boggy ground, for example. One of the most striking discoveries was of an adult mammoth skeleton near Naco, Arizona. It was found with eight Clovis points embedded in it, all thrust into critical areas. The fatal point may have been one driven into the base of the skull. It has been suggested that the Naco mammoth survived spearing and escaped its human predators, only to die later of its wounds.[4]

Although the Clovis people did seem to kill a disproportionate number of young mammoths, analyses of the creatures' age from kill-sites reinforces the idea that whole herds of mammoth were massacred—not just young, old or isolated individuals. In this the Clovis may have acted much like modern African elephant cullers who kill complete family units as the most efficient way to control population. Modern hunters usually kill the matriarch first, after which the other elephants mill about in confusion and are easily slaughtered. Detailed knowledge of the behaviour and herd structure of other large mammal species was used by Plains Indians and Inuit in their hunting, and there is no reason to believe that the Clovis people would have been ignorant of such methods. Such hunting obviously had the potential to lead to the rapid extinction of creatures such as elephants. Still, under favourable conditions elephant populations can increase by 7 per cent per year, so considerable hunting effort would have been required to drive them to extinction.[5]

All wild creatures need to learn what is dangerous and how best to avoid it. Creatures that flee too readily are disadvantaged because of the vast amount of energy they waste and the disruption this brings to their lives. The behaviours animals use to avoid predators are both genetically based and learned. The genetic component is acquired through natural selection and so can only slowly be developed. This may account in part for the fact that most of the world's surviving large mammals live in Africa, for it was there that humanity evolved, and it was only there that animals had the time to acquire the genetically based behaviours that allowed them to cope with the new predator.

African elephants have been evolving in the presence of fully modern humans for at least 100,000 years and as a result can be difficult to exterminate. The best example of this comes from the Addo region of South Africa, a place of dense bush and adjacent farmland. In 1919 a government hunter was ordered to exterminate a herd of elephants that was damaging orchards. The hunter quickly killed most of the group, but the few survivors became adept at avoiding him. They hid in close thickets by day, becoming nocturnal, and whenever he ventured into the dense bush he found that the tables were turning—they were stalking him! In the end he had to give up, and the remaining animals were later protected in the Addo Elephant Reserve. Even today, eighty years after their persecution, these elephants have a reputation for being the most dangerous in Africa.

There are good reasons to believe that the American mammoth was not as adept at avoiding hunters as the elephant survivors at Addo. This is because they had been evolving in North America for 1.7 million years, a continent from which humans and their relatives had been absent. Initially they would have had no fear response to humans at all, for an elephant that flees from a human-sized creature would spend all day running, and would uselessly expend much energy. Such a creature would have been selected against by hundreds of thousands of years of evolution. Instead, the appropriate response—of ignoring such small, inoffensive-looking creatures—would have been selected for genetically as well as in learned behaviour. Given the level of efficiency achieved by

Clovis hunters, it seems unlikely that the Columbian mammoth had time to acquire either an appropriate genetic or learned response to the threat humans posed.

The hunting of the mastodon is also attested to in the archaeological record. Mastodons were more stocky than mammoths and may have been less social, making them easier to hunt. New information about them and how they were hunted has come from excavations of various swamps and bogs. The Heisler site in Calhoun County, Michigan, is one such place. Thirteen thousand years ago it was an open pond a metre or two deep. When the site was excavated the bones of a young male mastodon were recovered. Some bones were associated with unusual features such as holes in the sediment that were infilled with clay from above, suggesting that poles had been driven into the pond around the bones. Strange strings of gravel also overlaid some bones. When these strings were examined under the microscope they were found to consist of gravel mixed with the stomach contents of a large herbivore.

These bizarre features of the site were interpreted by palaeontologist Dan Fisher to indicate a hitherto unknown meat storage method practised by Clovis hunters. The gravel had apparently been stuffed inside great sections of mastodon intestine, accounting for its mixing with bowel contents. These had then been laid atop piles of butchered mastodon meat in order to act as anchors. The poles may also have helped keep the meat submerged.[6]

There is a clear benefit to storing meat in chilled water—it won't rot as quickly as if it were left exposed to the air. If one is living with a whole megafauna there are other reasons one might want to do this, for dead bodies attract scavengers. Lions, giant short-faced bears, sabretooths and dire wolves, not to mention squadrons of feathered scavengers, would have quickly descended on an exposed kill, making it a very unpleasant camping location.

Evidence for such thorough dismemberment and storage of

mastodons has prompted researchers to experiment with elephant butchery themselves. In 1992 Dr R. S. Laub and seven assistants took the opportunity presented by a euthanased zoo elephant to investigate how Clovis people may have dismembered pachyderms. They soon discovered that their task was not an easy one. To begin the dissection Dr Laub took a well-sharpened, fifteen-centimetre-long fleshing knife and slashed at the skin of the deceased elephant's face. He reports (having apparently forgotten that pachyderm means 'thick skin' in Greek) that 'the knife proved surprisingly inadequate for the job. More than ten forceful sawing motions along the axis of the arch [cheekbone] were required to produce a small scratch.' At this point Dr Laub sensibly switched to using a surgical scalpel, for the skin resisted efforts to stab through it with anything else. Clearly not all options were tried, however, for Laub admits that consideration for the feelings of the zookeepers precluded anyone from taking a well-aimed blow at the much loved deceased proboscidean using their full bodyweight and a larger knife.[7]

Laub was impressed with the great weight of elephant skin, noting that a piece roughly a metre square 'could be moved by two men only with difficulty, by holding opposite edges and dragging it along the ground'. The experience, Laub commented, was like moving three or four oriental carpets stacked one upon the other. To compound the difficulty, the moistness of the hide made it quite slippery. Having defleshed one side, the team of eight attempted to turn the body over, requiring the services of a backhoe (presumably unavailable to Clovis hunters) to complete the task. As an exhausted Dr Laub ruefully noted, 'carrying out a defleshing operation such as ours in muddy or partly under water would be something to avoid at all costs'. Still, we modern humans can claim little expertise in the art of pachyderm disassembly; even butchering a sheep is beyond most of us.[8]

Professor Dan Fisher realised that if dead elephants could talk they could tell us what caused their demise, a sentiment that perhaps led this ingenious man to discover that even ancient elephant corpses can provide eloquent evidence of their lives and deaths. Fisher's work has involved the examination of numerous mammoth and mastodon tusks from Michigan and Ohio, the size and shape of which inform an experienced researcher of the sex of their owner. Furthermore, the microstructure of the tusk records many details of an individual's life. One can tell, for example, how many young a female gave birth to during her life, and at what intervals. This is because during the twenty-two months of her pregnancy the calcium that normally goes into growing a female's tusks is shunted into the growing bones of her baby, leaving distinct narrow bands in the tusk. The tusks can also be dated using the radiocarbon method.

Fisher collected the tusks of animals that lived about 13,000 years ago and tried to determine whether their owners were chronically starved, or were healthy and breeding rapidly. If they were starved, as many climate-based hypotheses suggest they should be, their tusks should show a pattern of stunted growth with no or few signs of reproduction in the undernourished females—the pattern seen in modern elephants during a prolonged drought. If, on the other hand, the mammoths were being overhunted, they should show signs of rapid growth due to ample nutrition (for there would be few mammoths and mastodons and lots of food), and the females should be having young frequently, just as modern elephants do in response to intense hunting. This happens because the young are often killed and there is plentiful food for the females, allowing for short spaces between births.

Although the sample size is still small, the tale told by Fisher's tusks is unequivocal. They were, by and large, from well-nourished individuals and the females were—by elephant standards—breeding furiously, producing a new young every four years. Only considerable predation, not a deteriorating climate, could account for such a pattern.[9]

You might judge from these studies that Dan Fisher has an ingenious if curious turn of mind. His most extraordinary discoveries, however, reveal just how agile that mind is. While excavating the now famous 'Burning Tree mastodon' (whose remains were unearthed during excavation of a pond on the Burning Tree golf course in Newark, Ohio) he encountered a sixty-centimetre-long, twelve-centimetre-wide mass of partly decomposed plant matter. It was reddish-brown in colour and had a 'pungent odour'. Suspecting that he had found the contents of a section of mastodon bowel, Fisher sent bacteria from it to be cultured. To his delight the agar plates inoculated with the material grew a variety of bacteria found in the gut of large herbivores, but very different from those living in the surrounding bog sediments. Incredible as it may seem, Fisher appears to have found organisms that last bred in the bowels of a mastodon. They had survived burial alive for nearly 13,000 years and had then begun to breed again on modern agar plates! Such a stunning discovery is still doubted by many, but there seems to be little *prima facie* reason to doubt the results.[10]

Fisher, curious as to why so many dismembered mastodons were cached in ponds by Clovis Indians, began his own experiments with caching meat. At first he used commercially butchered legs of lamb in Crane Pond and Big Cassandra Bog (both located near Hell, Michigan). Fisher set out the caches in autumn and left them for one to two years, checking some every two to four weeks but leaving others undisturbed. Bacteriological analysis of the cached meat showed that while it looked a bit grim it had bacterial levels as low as those of meat stored in commercial freezers. Clearly, caches of mastodon meat could be visited and utilised for years after a kill had been made, yet Fisher was unsatisfied as to whether lamb legs were an adequate substitute for diverse cuts of mastodon.

In February 1993, when a friend's 680-kilogram draughthorse died he got the chance to experiment further. Fisher skinned and broke up the carcass then cached the meat. He also made 'anchors' for the meat from gravel-filled sections of intestine, just as he believed the Clovis Indians had done. These intestinal anchors surprised him, for they began acting in odd ways. To begin, gas accumulated at one end of the tied-off intestine. This end then floated to the surface and remained there until 'they finally decomposed in August'. Right through winter the gassy gut-ends performed curious bobbing motions in the water, keeping a small area of the pond from freezing over. Such anchors were doubtless useful to prehistoric hunters, both in marking the location of the meat stash and in facilitating retrieval. Fisher wasn't the only one interested in the guts, for footprints revealed that 'small animals' took an interest in these floating tidbits as well.[11]

Fisher checked on the horsemeat every two weeks. In March he reported that the meat 'took on a progressively stronger "cheesy" smell and taste'. Worse was to come. By April, portions of the dubious dinner had broken loose and were found floating in the pond where they grew a profuse coat of green algae. Some bits even floated to shore, where unidentified canids made a meal of them. Pursuing his experiment with exceptional dedication, Fisher retrieved the green hunks of horseflesh and re-tethered them to the bottom. By June, he noted ominously, 'the meat and fat remained edible'. At first I assumed that Fisher was referring to its gastronomic acceptability to the unidentified canids, but then I read on and discovered that 'despite a strong smell (subdued by cooking) and sour taste, the meat retained considerable nutritive value until July and August, when advanced tissue breakdown led to bone exposure and incipient disarticulation'. One wonders whether 1993 saw the Fisher family sitting round a fortnightly roast of horsemeat that had been cached by Dad in the pond near Hell. Academic salaries are notoriously slight, but even so Fisher's devotion to science was clearly above and beyond the call of duty. Were there a Nobel Prize for prehistory I would award it to him, for through his gastronomic intrepidness a major contribution to prehistory was made. He proved that caching meat in ponds allowed for its use many months after the kill had been made, and when one is faced with consuming a mastodon, such know-how is valuable indeed.[12]

Although the remains of horses and a few similar-sized species have been found at some kill-sites, evidence that humans hunted other North American megafauna is rare. The only species for which convincing multiple Clovis kill-sites exist are the mammoth, mastodon and giant bison. Just why kill-sites have survived for these species and not for others is puzzling until we consider the fact that these resilient creatures lasted a little longer than the others (the bison eventually transforming into the modern kind) and that this may have increased the odds that kill-sites containing their remains would be preserved.

The discovery that the mammoth and mastodon outlasted other megafaunal species puts paid to another innovative idea about why America's megafauna died out. This was based on the theory that very large animals such as elephants are keystone species in the environment. If they were killed out by hunters, it was argued, then maybe the other, smaller kinds of megafauna followed in an extinction 'cascade' as the environment changed. The fact that elephants apparently became extinct after everything else disproves that this happened in North America south of the ice, although the idea may still apply for the Arctic, where the special conditions of the mammoth steppe may have permitted such a cascade.[13]

Around the time the last American elephants died, approximately 12,800 to 12,900 years ago, people ceased manufacturing Clovis points. Although various stone points continued to be produced for thousands of years, Clovis by then was evidently a superannuated technology. If our current chronology is accurate and humans were indeed the cause of the extinction of America's megafauna, it had taken just 300 years to dispatch into oblivion, through the black hole that lay between Clovis nose and Clovis chin, a continent full of giants.

The Uinta beast, *Uintatherium*. The largest creature of its day, this nightmare of an animal thrived until 40 million years ago.

Titanis: a predatory bird of 400 kilograms that stalked Florida and Central America until 13,000 years ago. The group originated in South America.

A gigantic enteledont feasts upon the carcass of an early rhinoceros. Skulls found of the largest enteledonts are up to a metre long.

A short-faced giant sloth, *Glossotherium* or the 'tongue-beast' as its name translates. It flourished in North America until 13,000 years ago.

The American mastodon (*Mammut americanus*) survived in North America's eastern forests until 13,000 years ago, long after the extinction of mastodons elsewhere.

Thirteen-thousand-year-old Clovis points and other tools—signature pieces of northern North America's first pioneers.

The American condor is a last, endangered relic of North America's ice-age megafauna.

The 1979 discovery of this 35,000-year-old long-horned bison carcass in Alaska revolutionised ideas about bison evolution. The mummy was dubbed Blue Babe because of the blue vivianite encrusting its skin.

The snapping turtles are some of North America's most venerable (if ugly) inhabitants. They have lived on the continent for 70 million years.

Amerigo Vespucci: 'The pickle dealer at Seville... who managed in this lying world to supplant Columbus and baptize half the earth with his own dishonest name.' Ralph Waldo Emerson

George McJunkin: cowboy, astronomer and archaeologist. His discoveries opened a window on the deep prehistoric past of Americans.

Ishi: the last Indian to escape the Europeans in the US, who was starving when found in 1911. This photo was taken around 1915, a year before he died.

Geronimo: one of the last native Americans to deny the power of the US military.

Emigrants Crossing the Plains. Albert Bierstadt gives the nineteenth-century conquest of the west a golden glow.

After the last wild herds were destroyed, people turned to buffalo bones for a living. Here skulls are stockpiled at Detroit, Michigan, before being freighted to a carbon works.

Regardless of whether human hunters or climate change caused the extinctions, the event is without parallel in North American prehistory; for no earlier extinction event has been correlated with the arrival of a new species, or with a warming climate. None so specifically targeted the large land mammals, and none except the asteroid-caused extinction of 65 million years ago left such an impoverished ecosystem. This was a condition the continent had never existed in before; its large creatures had been banished, yet its smaller denizens remained largely unaffected by extinction.

It is now time to move on from this discussion of the causes of the extinction, and to consider how North America reconstituted itself ecologically following this signal event.

Sixteen

THE NEW AMERICAN FAUNA

The North American environment changed dramatically around the time the megafauna vanished. Most researchers attribute the changes to the same climatic factors they believe caused the extinction of the great mammals. While there is no doubt climate played a role in the transformation of ecosystems, I think we must look elsewhere if we are to understand the changes fully, and one of the best places to start is the effect the great mammals themselves had on the landscape.

The most dramatic ecosystem change to occur 13,000 years ago took place north of the North American ice sheet, in Alaska. There, an entire plant community vanished, along with the mega-mammals it fed. Dr Dale Guthrie of the University of Alaska has spent a lifetime studying this vanished ecosystem. He argues strenuously that when the mammoth lived, Alaska and Beringia supported a very different kind of vegetation which, despite the intense cold of the glacial maximum, was much more productive than today's. Guthrie has called this vanished vegetation the mammoth steppe.[1]

As Guthrie reconstructs it, the mammoth steppe was a grassland with lots of bare ground between the tussocks. This Alaskan vegetation arose, he says, because extreme dryness inhibited the growth of trees, shrubs and many other plants. There is certainly much evidence for drier conditions during the glacial maximum in the form of windblown dust, called loess. Guthrie thinks that the bare earth between the grasses allowed the ground to thaw earlier and freeze later, providing a longer growing season. He points out that nutrient recycling would have been highly efficient in such an ecosystem because large animals would have composted the grass quickly via their dung.

Today's Alaskan vegetation is very different from that of the hypothesised mammoth steppe. Now, a layer of acidic, undecomposed plant matter builds up under the living woody vegetation. This allows the permafrost to creep up from below and prevents nutrients getting into or out of the frozen soil. Soon the vegetation turns 'sour' or un-nutritious for large mammals. In these conditions creatures such as bison (which were reintroduced to Alaska in historic times) find it difficult to get enough good-quality fodder to survive.

Some palaeobotanists dispute whether the mammoth steppe ever really existed. They suggest that even under the coldest conditions the Bering land bridge provides no fossil evidence that there ever was a mammoth steppe, but that instead a wetter grassy tundra with birch flourished there. Despite such objections, it seems to me that Guthrie's arguments are on the whole sound, for clearly Beringia (including Alaska) supported a considerable biomass of large animals, something that it cannot do today. Perhaps the mammoth steppe was restricted to drier environments and thus left little fossil record.[2]

Where I disagree with Guthrie is on the reason for the demise of the mammoth steppe. He thinks that its disappearance was due to climate change, and that this in turn led to the extinction of the mammoth and other megafauna. There is one clear weakness in this argument. The climate has warmed seventeen times in the last 2.4 million years, just as it did when the mammoths vanished 13,000 years ago. Yet the fossil record indicates that throughout these earlier times Alaska supported a megafauna which included the mammoth after its arrival 1.7 million years ago, and that the warming did not bring extinction to either mammoth or mammoth steppe. Why should just the last warming event of seventeen bring disaster? It is necessary to find a new factor not present during earlier warming events to explain this.

Guthrie and others argue that the nature of the last warming event was somehow unique, though evidence for this notion is rather scant. A more convincing explanation is that the new factor was man the hunter, and that the sequence of extinctions was the other way around— the extinction of the mammoth led to the destruction of the mammoth steppe. My reasoning is based on the idea that elephants and other truly large mammals (those weighing over a tonne) are profound modifiers of their environment. In Africa, elephants can convert forest into savanna through their destructive feeding on trees, thereby creating much usable habitat for grazing species. On the mammoth steppe a ten-tonne mammoth would have consumed between 100 and 120 kilograms of vegetation (much of it possibly woody) each day, dumping out waste in huge bowel movements at almost hourly intervals. The effect of such large herbivores is profound in regions, such as Alaska, where cold or drought reduces levels of primary productivity.[3]

What happened when the mammoth finally vanished from Alaska? Doubtless much coarse vegetation which otherwise would have been eaten was left untouched. Whatever lesser megafauna remained, such as caribou, would not have been able to consume this coarse herbage because only the largest herbivores can profit from such poor-quality food. This vegetation would have eventually built up into insulating mats that inhibit nutrient flow and support the inedible vegetation that occupies Alaska today.

A warming climate would have hastened the process, accelerating the growth of woody plants and, paradoxically, due to increased evaporation from warmer oceans, bringing more snow. The result of mammoth extinction under these circumstances might have been doubly disastrous, for there is good evidence in the form of worn tusks that these titans removed snow from grasses in winter, making food accessible to smaller species such as horse and bison. Extinction under such circumstances would have been a terrible blow to the whole ecosystem—horses, bison and other grazers would have starved during the winter, while the thaw would have brought greater growth of woody, inedible and insulating plants. Alaska would have thus rapidly turned from a herbivore-friendly to a herbivore-hostile place, the various species vanishing as a cascade swept through the ecosystem, beginning with mammoth, then horse, bison and mammoth steppe (with its still poorly understood assemblage of invertebrates and plants).

Below the Arctic Circle, disruption to the ecosystem following the megafaunal extinction was quite different. I would argue that it has been misunderstood in much the same way as has the demise of the mammoth steppe. In North America south of the ice, two major changes occurred around the time of megafaunal extinction: the distribution of many small and middle-sized mammals contracted, and a new megafauna invaded.

The phenomenon of range retraction among many of North America's smaller mammal species at the time of megafaunal extinction has long been recognised. Researchers have characterised the phenomenon by studying small mammal communities rather than individual species, and have developed an ingenious model to express their ideas. As the distributions of various small mammal species retracted they no longer overlapped, and as a result many small mammal communities became less diverse. Dr Russ Graham and Professor Ernest Lundelius describe these more diverse small mammal communities as 'disharmonious', presumably because the mammals then living together disturb their concept of ecosystem harmony based on understandings of the world as it is today. The most marked examples of these communities are found on the southern Great Plains. There, 14,000 years ago, northern species such as the short-tailed shrew and Richardson's ground squirrel lived with southern species such as the cotton rat and giant armadillo, as well as eastern species such as the eastern chipmunk and Carolina shrew. Graham and Lundelius hypothesise that these rich 'disharmonious' small mammal communities existed because the climate then had more modest seasonal extremes, which allowed more species to co-exist.[4]

There is another possible explanation, however, for the extinction of the megafauna could also have led to the observed range retraction of the small mammals. As elephant, camel, peccary and ground sloth were eliminated, disturbances created by knocking down trees and digging wallows would have been much reduced. The vegetation may have responded to this lack of disturbance by becoming more uniform—an eternal gloomy forest without breaks, or a uniform grassland without wallows, herb patches and shrubs. These more uniform environments would have offered fewer opportunities for the small mammals to find a niche. Various species would then have retreated to their core regions, resulting in the distribution patterns seen today. To me it seems more appropriate to think of today's mammal communities as the 'disharmonious' ones, for they seem out of step with what has existed for most of the last two million years.

As with the case of the mammoth steppe, a test of these ideas can be made by examining what happened to small mammal distributions during the sixteen warming episodes before 13,000 years ago. Data about small mammal assemblages during these earlier times is still scarce, but an interesting fauna, dating to a warm phase about 425,000 years ago, has been found in Porcupine Cave, high in the mountains of Colorado.[5] There is a greater variety of rodents in the Porcupine Cave fauna than can be found at present in any similar North American environment. The fact that high diversity occurred during a warm period in the past but does not occur in today's conditions suggests that small mammal diversity was responding not so much to climate as to some other factor, possibly the greater environmental diversity created by megafauna.

It was not only small mammals that retracted in distribution during this period, for a number of middle-sized species withdrew from much of the country and, amazingly, they all withdrew southwards, many not stopping until they reached the jungles and thickets of what is now Mexico. Here I see a different cause for their southern migration—direct hunting by humans.

Beginning around 13,000 years ago, species as diverse as armadillos, tapirs, jaguars, spectacled bears, llama, ocellated turkeys and peccaries all moved southwards. This is quite a surprising pattern, for all of these warmth-loving species were withdrawing from the north of the continent just as it was heating up. Just why they survived in South or Central America while becoming extinct in the northern margins of their range is therefore an intriguing question. Some insights may be gained from the history of the jaguar.

Jaguar roamed the boreal forests south of the ice sheets 14,000 years ago, much as a few leopards still do in Siberia today. As late as the eighteenth century they survived as far north as the Tehachapi Mountains in south-eastern California, the Grand Canyon, and into the Appalachians. Hunting by a carnivore-phobic European population eliminated the species north of Mexico and the last jaguar killed in the United States was shot in Texas in 1946.[6] It is obvious that hunting, not climate, is responsible for this reduction, as I suspect it was for all the prehistoric ones. The jaguar, interestingly, is just beginning to be seen in the United States again, but now it is being photographed rather than shot. If such circumstances continue, I predict a rapid northwards expansion of its population that may eventually extend as far as southern Canada.

Tapirs, jaguars, llama and spectacled bears are all large animals that have found refuge in the dense jungles and broken terrain of South America. Dense rainforest is a particularly difficult habitat for humans, and it along with swamps, steep forest-covered mountainsides or high and desolate plains form the present refuges of these creatures. Their populations in North America may have always been rather marginal (although their extensive fossil record there belies this), readily affected by disruption of migration or straight-out hunting in accessible habitats. Nevertheless, their survival in the south, often outside the bounds of the United States, has caused Americans to think of them as alien fauna. In reality they are indigenes whose reintroduction could benefit many American ecosystems.

While the small and medium-sized mammals were retreating, a few large mammal survivors of the Pleistocene blitzkrieg were quietly expanding their ranges. All of these survivors originated north of the ice sheet in Beringia and, just as countless species had done before them, they were heading south into the American heartland. Such was their success that a visitor to Yellowstone National Park and other such reserves in North America could be forgiven for believing that the continent had never suffered megafaunal extinction. Great herds of bison and elk, the stately moose, the formidable grizzly and the wolf still stalk a landscape seemingly replete with large mammals. This assemblage does indeed represent a true if limited megafauna, and the story of how it came to inhabit North America is one of the most remarkable sagas in the entire history of the continent.

Consider these creatures by another set of names and their origins become clear. Bison is just another name for European wisent (*Bison bison*), elk for red deer (*Cervus elaphus*) and grizzly for brown bear (*Ursus arctos*). All of these creatures are very recent Eurasian immigrants to North America south of Alaska, so recent in fact that they either entered the continent at the same time or just a little after the Clovis people 13,000 years ago or, if present earlier, have risen to dominance since that time.

The very first bison like-creatures entered North America from Eurasia about 400,000 years ago and, ever since, periods of isolation have alternated with periods of renewed migration of bison from Europe. By 36,000 years ago a gigantic, long-horned species had evolved. There were two main populations, one roaming the American prairie south of the ice sheets, the other inhabiting a vast stretch of land from Europe then east to ice-free Alaska. These populations, which many researchers classify as different species, were similar in their physiognomy but differed in one vital respect. The Eurasian–Alaskan population (one continuous genetic population that straddled the Bering land bridge), had evolved in the presence of human hunters, at least on its eastern and southern margins. Some of the behaviours that allow creatures to avoid predators are 'hard wired'. They are genetically controlled, and the Eurasian–Alaskan bison had been selected over tens of thousands of years to behave in ways that allowed them to survive human hunting.

The bison living south of the Laurentide ice sheet, however, had become isolated in North America before humans had settled north of 50 degrees in Eurasia. Thus they almost certainly lacked the behaviours that I think allowed the Eurasian–Alaskan bison to survive hunting. In any event they quickly became extinct when Clovis hunters arrived south of the ice about 13,000 years ago. North America was not to remain long without bison, however, for as the ice sheets melted the Eurasian–Alaskan bison migrated south, inheriting the vast prairies not long vacated by their American brethren.[7]

Until 13,000 years ago no grizzly bear ever prowled North America south of the ice, for the species only reached Alaska from Eurasia 50,000 to 70,000 years ago. Instead the continent was home to a more horrible predator, the gigantic short-faced bear, which was thirty-odd centimetres taller at the shoulder and twice the weight of a Kodiak bear. It was a swift runner and a specialised meat-eater capable of dispatching any living creature except perhaps a healthy adult mammoth or mastodon. Just how Clovis people coped with this terrifying creature no one knows. The bears, however, would have had no experience or fear of humans, and may have viewed them as very slow-moving prey. As a consequence, many probably ended their days impaled upon Clovis points.[8]

Tom Stafford has uncovered evidence that short-faced bears might have hung on for a little longer than other megafauna, surviving until 11,500 years ago on the southern prairie. Stafford and Russ Graham suspect that the bears made a marginal living hunting buffalo and were mostly avoided by human hunters. If so, the great carnivores were certainly relict by then—just as condors are today—and their extinction left the large-bear niche open in North America. The more versatile omnivorous brown bear or grizzly, with its long experience of human hunters, soon filled the continent in its wake. Unlike bison, no brown bears of any sort had inhabited North America before. Although brown and short-faced bears may have co-existed in Alaska (the fossil record is equivocal), it seems possible that competition from short-faced bears excluded the browns from North America south of the ice.

The story of the moose (*Alces alces*) is one of the best documented of any of the new immigrants. It evolved in northern Eurasia from smaller, more typical deer-like relatives about two million years ago. Early moose, known as stag-moose, spread to Arctic North America some time between 800,000 and 200,000 years ago, possibly splitting into distinct eastern and western populations. Both became separated from their Eurasian and Alaskan relatives, which continued to evolve in a separate direction, giving rise to the modern moose by about 100,000 years ago. About 13,000 years ago the stag-moose was exterminated along with the rest of North America's megafauna. Soon after, the continent south of the ice was invaded by the Eurasian moose, which spread rapidly to occupy all suitable habitats.[9]

The stag-moose illustrates a curious fact often seen in the fossil record of North America—that the continent tends to slow the evolution of some of its immigrants. By the time it became extinct it was a relict, a living fossil similar to the moose that inhabited Europe around half a million years earlier; it is as if its evolution had stood still while that of its Eurasian brethren robustly continued. The jaguar provides another example of this phenomenon. The bones of living jaguar display similarities to million-year-old ancestral jaguar/leopard fossils from Europe, but differ from those of living leopards.[10] The Afro-Eurasian leopard appears to have continued to evolve physically, while the evolution of the American jaguar slowed, so that it still resembles its ancestor. In respect to both the moose and jaguar, the stay-at-home European populations gave rise to different-looking species (modern moose and leopard) while their immigrant American cousins remained similar to their fossil ancestors. In a related phenomenon, North America has often acted as a last refuge for archaic groups such as multituberculates, mastodons and sabre-tooths, long after their extinction elsewhere.

This pattern of survival of ancient forms in the Americas is seen in several other North American species, but just how widespread it is remains to be determined. Why it occurs at all is also far from clear, but part of the answer appears to lie in the amount of competition a species faces. All other things being equal, larger landmasses possess more species than do smaller ones, so competition and thus evolutionary change are faster there. Such a theory predicts that North America, which is third in size among the continents, should be full of creatures that, on average, exhibit slower rates of evolutionary change than those of Eurasia and Africa. Australia is the smallest of the continents, and its fauna seems to support this idea, for it is famous for being the home of many 'living fossils' like the platypus, which have changed little over millions of years.

The story of the grey wolf (*Canis lupus*) differs from those of the moose and jaguar. It appears to have first evolved in the Arctic biome, which includes the high latitude parts of Eurasia and Alaska. By about half a million years ago it had moved south into Eurasia, becoming almost indistinguishable from the modern wolf. In the meantime, North America had become populated with a larger kind of wolf, known as the dire wolf (*Canis dirus*), which had originated in North America south of the ice from earlier wolf-like creatures.

At some time over the past 130,000 years grey wolves arrived in North America south of the glaciers. Some scientists think that they were around 100,000 years ago and found a niche for themselves as a middle-sized canid, squeezed between those occupied by the dire wolf and coyote.[11] Other researchers, however, believe that they are extremely recent immigrants, having arrived less than 13,000 years ago—after the extinction of dire wolves and the arrival of people. Whatever the case, grey wolves only become common in the North American fossil record less than 13,000 years ago.[12]

The origins of the puma, North America's remaining large carnivore, is also turning out to be rather strange. Recent DNA analysis has revealed that the continent's pumas have passed through a genetic 'bottleneck' some time over the last 13,000 years. Some think that this occurred when the North American puma became extinct and the species re-invaded the continent from South America via the Isthmus of Darien.[13]

The history of the elk (red deer) is still shrouded in mystery. It was long believed to have been resident in North America for a million or more years, but some researchers now think that it too arrived around 13,000 years ago. This reassessment has arisen from a critical examination of fossils thought to represent elk from earlier periods. All are very fragmentary and difficult to distinguish from the remains of other deer. The most sceptical scientists dismiss the fossils as belonging to various species of living and extinct American deer such as mule deer, while others maintain that at least a few of the fossils represent elk. The earliest universally accepted evidence of elk fossils in North America found south of the ice appears to be about 13,000 years old. If indeed they are such late arrivals, it is not clear what could have held them back, but it is possible that they were excluded from North America by competition with any number of now-extinct browsing species.[14]

The high Arctic of the New World served as a refuge for the musk ox (*Ovibos moschatus*), long after it became extinct in the old. Sophisticated hunters apparently drove this strange creature to extinction in Eurasia 3000 years ago, but in the New World Inuit hunted them only in times of desperation, saving them as a larder of last resort. This was in effect a conservation strategy that might have saved these magnificent creatures from total extinction. Their survival in the New World Arctic has allowed them to be reintroduced to northern Eurasia in historic times, where they are now a valued element of the tundra biome. Other tundra species such as caribou (reindeer) and polar bear represent a circumpolar fauna that, whatever its origins, is neither American nor Eurasian in primary distribution. Somehow they have managed to adapt to the human presence and survive today throughout much of their former range.[15]

Setting aside the mystery of the musk ox and the uniformity of the Arctic fauna, the general pattern exhibited by the largest creatures of North America is clear. Most, if not all, represent a new megafauna, arrived south of the ice from Eurasia and Alaska. The largest native Americans clearly to survive the extinctions south of the ice are the black bear, which arrived around one million years ago, and mule deer. Neither is much bigger than a human, nor much larger than the largest surviving mammals of the other extinction-blighted continents of Australia and South America.

North America's new megafauna dramatically changed the continent, and it is in the crucible of that change that a distinctive American ecosystem was forged. The America of Geronimo and Buffalo Bill would be assembled in a breathtaking evolutionary whirlwind of change that acted for just thirteen millennia.

SEVENTEEN

THE MAKING OF THE BUFFALO

With the end of the Clovis culture about 12,900 years ago, America's first human frontier closed. The first pan-North American culture had vanished and the varied Indian cultures to which it gave rise would live within the bounds of a diverse yet impoverished continent. Of all the great beasts that fed the Clovis people, just one was still present—the American buffalo or bison—but even it was now represented by a different genetic stock and had suffered a diminution in range, for by this time it had become restricted to the great prairie of the American west. By about 12,950 years ago a group of Indians had begun to manufacture a new kind of stone point in order to exploit this resource, and in so doing would change the buffalo forever.

When President Grover Cleveland wed Frances Folsom in 1886, he unwittingly changed the way that we would think of America's first post-Clovis culture. Cleveland's wife had lived for a time in Ragtown, New Mexico. The town had been founded in 1858 as a railroad construction camp and in 1888 its citizens were so overwhelmed at the good fortune of their girl next door that they changed its name in her honour. Folsom it was from then on, which would mean little to anyone if it had not been for a remarkable African-American cowboy who settled in the area, and the discoveries he made.

George McJunkin was one of those people who seemed to be curious about everything. A childhood spent in slavery had deprived him of an education, but after being freed by Union forces in 1865 at the age of fourteen, he took up the life of a cowboy on the Crowfoot Ranch near Ragtown. There, the indomitable George got the sons of a rancher to teach him to read. He eventually became a much sought-after fiddler, an amateur surveyor, astronomer, maker of survey instruments and in time the respected foreman of the ranch. Notwithstanding these achievements, George's real passion lay in the study of natural history. Wherever he went he picked up curios that he kept in a small museum atop his fireplace. There, stone arrowheads competed for space with fossilised animal bones and the skull of a prehistoric Indian.[1]

In 1908 McJunkin was out breaking horses at a place he called the Wild Horse Arroyo when he spotted a large bone protruding from the side of an erosion gully. He dismounted and dug carefully around the bone, only to find that it was attached to another. McJunkin knew that he had stumbled across something important, recognising the bones as belonging to a very large bison-like creature. He returned so often to excavate his finds that the place became known as 'McJunkin's bone pit'. In an effort to convince others of the importance of the discovery, McJunkin wrote letters to friends describing it. When he died in 1922, before a professional excavation could be organised, his friends gave him a cowboy's send-off, lowering him into his grave on their lariats. They also remembered the discovery and in 1926, along with Jesse Figgins, Harold Cook and Marie Wormington from the Colorado Museum of Natural History, finally carried out the long-planned dig. What they found astounded them and revolutionised our understanding of American prehistory.

As the team dug they uncovered the near-perfect skeletons of several giant bison. This was a spectacular enough find in itself, but what really put the place on the map was the discovery of a meticulously crafted stone point, once used to tip a wooden spear, between the ribs of one of the fallen giants. Here was evidence that the great, long-horned bison entombed in McJunkin's bone pit had been felled by the hand of man.

In 1926 it was widely believed that Indians had been resident in North America for just a few thousand years. The discovery at first generated disbelief at the co-occurrence of a spear point with the bones of extinct creatures, but further excavation of the site, attended by some of America's most sceptical scientists, convinced even the worst doubting Thomas of the authenticity of the association. At one stroke of the trowel North America had been granted a deep human prehistory and the down-at-heel town of Folsom gave its name to an entire stone-tool tradition and a long-vanished culture.[2]

We now know that Folsom-style stone points were made over a period of 700 years, from between about 12,950 and 12,250 years ago. It is obvious that their manufacture was a labour of love, for few people anywhere have ever crafted such elegant artefacts from stone. The points are closely worked to give them a sharp edge, and are deeply fluted on their sides. Fluting involves striking a long flake off each side of the flattened blade, thinning it considerably and giving it concave rather than convex surfaces. It is a risky operation, for the point is liable to break as the flake is struck off. It has the virtue, however, of producing a more penetrating point whose concave sides allow a wound to bleed freely. Such points are ideal for quickly killing the giant long-horned bison which, being more fleet than an elephant, can run many miles before expiring unless severely weakened. George McJunkin's bone pit revealed that some 12,900 years ago, twenty-three long-horned bison met their deaths via the tips of several dozen of these points.[3]

The transition from Clovis to Folsom is important because in it we see one of the first regional cultural differentiations in North America. Folsom was centred in the west, where the buffalo roam.

Folsom soon gave way to a simpler variation in that these new points were not fluted. By 12,000 years ago the use of spears to hunt bison was perhaps becoming obsolete, for mass kills often involving hundreds of animals reveal that Plano people (those who made the newer points) had begun to drive-hunt rather than spear the great beasts. With large amounts of meat becoming available to these smallish groups there was inevitable waste. At the Olsen-Chubbuck site in eastern Colorado, for example, nineteen of the 193 bison killed were not even touched by the hunters while another thirty were only partially butchered. This waste, however, seems minimal when compared with that of nineteenth-century Europeans, who slaughtered bison merely for their tongues or skins, with just one skin making it to market for every five animals slaughtered.[4]

Although the Folsom and Plano people hunted a wide variety of animals including antelope, elk, deer and coyote and also harvested a variety of plants, excavation of their campsites shows that long-horned bison was their principal prey. The survival of an entire group must have frequently depended upon the success of a bison hunt, and Folsom culture was doubtless shaped by bison just as plains Indian cultures were in historic times. It is equally true, however, that Indian cultures shaped the bison. The proof of this is in the fossil record, for it tells us that bison were very different creatures before humans disturbed the New World.[5]

Dale Guthrie of the University of Alaska is perhaps the foremost living expert on the evolution of the bison, both fossil and modern. In July 1979 fate handed him a key that would permit him to fully understand how bison had changed over time. In that month a gold miner working north of Fairbanks, Alaska, called the university to say that he had uncovered the remains of a large creature. The find, made just as Guthrie was about to go on sabbatical to Europe, turned out to be a complete, albeit scavenged, mummy of an extinct kind of bison known as *Bison priscus.*

The bull had been felled in his prime, and his appearance was striking, for over the thousands of years he had lain buried a coating of vivianite crystals had stained his skin bright blue. When Guthrie saw the beast he was reminded of the Disney stories about Paul Bunyan and Babe, a legendary blue ox that accompanied Bunyan on his adventures. Guthrie's own Blue Babe—as he dubbed his find—was to lead the palaeontologist on an adventure back through 36,000 years of four-dimensional bio-space to reveal an almost unimaginable saga of bison history.[6]

Analysing the mummified and frozen remains—which included red meat, fat, hair and some organs—was to occupy Guthrie for years. In time his painstaking study of this unique specimen would reveal in enormous detail how bison appeared, lived and died 36,000 years ago. Most importantly, Guthrie argues, comparisons with living bison show just how much they have changed; as an Alaskan bison Blue Babe was drawn from the population that would give rise to the living North American plains and woodland bison.

Blue Babe lived at a time when mammoths, horses, bison and caribou were the main herbivores of the Alaskan tundra. The now extinct stag-moose and the living musk ox were also present, but in smaller numbers. These creatures formed the prey of just two principal predators—wolves and lions. At first Guthrie suspected that wolves had killed Blue Babe. After an exhaustive autopsy, however, he was able to determine that it was lions that had pulled down the 800-kilogram bison bull. They had dispatched him by suffocation, one placing its mouth over his muzzle (in the process leaving much bruising), just as African lions do with Cape buffalo today. One lion had even left a sizeable portion of its carnassial (meat-slicing) tooth in Blue Babe's frozen hide, presumably when it came back to try to feed on the frozen carcass.

Blue Babe had just enough meat taken off his bones to feed a few hungry lions, leading Guthrie to deduce that Alaskan lions lived and hunted in small prides—perhaps containing as few as two or three. As befits such unsociable creatures, the males probably lacked the huge dark manes used by their African counterparts to frighten off rivals intent on poaching their harems. Instead, Alaskan lions resembled the almost maneless lions that still survive in the Gir Forest of India, as well as those depicted in European cave art of 20,000 years ago.

Blue Babe was killed at the beginning of winter and his body quickly froze. Next spring it was covered by sediment that was washed downslope by heavy rainfall, quickly burying the body and preventing it from rotting in summer. The preservation of the corpse was so good in fact that, to celebrate the end of his studies, Guthrie invited a few colleagues to share a stew made from part of Blue Babe's neck. He went on to record that 'the meat was well aged but still a little tough, and it gave the stew a strong Pleistocene aroma, but nobody there would have dared miss it'.[7]

The most noticeable difference between Blue Babe and living bison were his horns. These were enormous by plains bison (*Bison bison bison*) standards, and stuck out rather than curving sharply upward, a characteristic marking Blue Babe as a member of a species known as the giant long-horned bison (*Bison priscus*). Although nothing like *Bison priscus* exists today, Guthrie discovered that it is not extinct, just irrevocably transformed by evolution into something else.

Blue Babe, Guthrie says, looked like a dark bay horse; a rich dark brown with black points on the face, legs and tail. This is very different from the typical colouration of plains bison bulls, which have jet-black forequarters and a contrasting sandy-coloured hump. Blue Babe also lacks 'pantaloons'—those long tufts of fur that give the forelegs of plains bison such a distinctive appearance—though he is significantly larger, as well as larger relative to his cows, than a contemporary plains bison bull.

More curious differences emerged as Guthrie began to study Blue Babe's bones. His hump (which unfortunately had been mostly eaten away by hungry lions) was different in shape from that of plains bison, suggesting that his head was held higher off the ground. The arc of Blue Babe's lower incisor arcade was also narrower, indicating that he took less in with each bite than does a plains bison. These differences may seem to be disparate in nature, even random perhaps, but Guthrie showed that they all stem from a single cause—a tendency for bison to congregate in larger herds.

One of the most telling differences of this shift is the change in the relative size of bulls and cows. A discrepancy in size between the sexes in grazing animals often occurs when males and females forage separately for part or most of the year. Blue Babe, Guthrie has suggested, could have grown so much larger than his cows only if he spent the growing season feeding alone in the areas of best forage, which he must have defended from all comers. Plains bison bulls are not that much larger than their cows because they must feed in competition with the thousands, if not millions, of individuals of both sexes around them. This restricts their access to good food and thus limits their size.

The drab colour of Blue Babe's coat and his absence of pantaloons also indicate that he foraged away from his females. He did not need to stand out in colour because his large size and his location indicated that he was a dominant bull capable of finding and defending the best territory. Plains bison bulls face a much more difficult task attracting females because they are just one not-very-large animal in an enormous crowd. The key to their dilemma, it turns out, lies in advertising by way of colour and display. Plains bison bulls differ from cows in their enormous black forequarters (emphasised by their pantaloons) and their contrasting sandy hump. These features advertise a bull's masculinity and strength, in effect saying to the cows—look what a fine beast I am!

It was Blue Babe's long horns, however, that provided conclusive evidence of an ecological shift. Today, plains bison bulls compete by engaging in the spectacular sport of head-butting. They run towards each other at full speed, clashing head-on with a loud and violent shock. This spectacular tactic is guaranteed to grab a female's attention, even in a crowd, for the commotion is audible more than a kilometre away. Had Blue Babe used his horns for head-butting they would have been broken by the shock. Short, curled horns are less vulnerable and far more effective as battering rams.

This difference led Guthrie to propose that Blue Babe used the less spectacular but equally deadly method of pushing to overcome rival bulls, just as cattle do today. In such contests the combatants look for a vulnerable area in which to thrust a sharp horn tip. Long, outward-directed horns provide a marked advantage because they can skirt round the protective horns of an opponent, thrusting deep into the shoulder. Guthrie has found fossil shoulder bones of long-horned bison bearing holes that neatly fit Blue Babe's horn tips, which supports his idea. Their bearers may well have died in combat, for the injuries show no sign of healing.

The plains bison's hump shape, lowered head and incisor width also indicate adjustments to life in the crowd. Plains bison occurred in such vast numbers that they cropped the grass into short 'buffalo lawns'. Such lawns are a nutritious source of food but each mouthful is competed for by the hundreds of other giant lawnmowers in the herd. For a plains bison to be successful in feeding it has to keep its head to the ground, taking broad bites that yield lots of food with each swipe. Their high hump aids in this, for it is in effect a cantilever that saves energy by ensuring that the 'at rest' position of the head is near the ground. In comparison, Blue Babe would have held his head high while at rest, scanning the horizon for predators trying to sneak up on the solitary animal. When he ate, his narrower incisor region allowed him to be more selective, plucking only the most nutritious of plants.

Life in a herd has many disadvantages. In addition to increased competition for food and a mate the chance of contracting gut parasites is increased, as is that of catching any infectious disease. If life in a herd is so challenging, why did the plains bison adopt the lifestyle so dramatically over such a brief period of time? The answer lies, it seems, in the dangers that plains bison faced. Herd behaviour is often a response to the presence of predators that can easily pick off isolated individuals or small groups. Such predators find it much more difficult to kill one of a herd, where the vigilance of a thousand eyes and the horns of companions affords protection.

The Great Plains and its contiguous grassed environments form the largest grassland in the world. It is a difficult place to hide, so it comes as no surprise to learn that the herding behaviour of buffalo was developed to its apogee there. By the time the plains bison inherited this realm, however, most of the great predators—sabretooth, lion, short-faced bear and dire wolf—were gone. There were only three predators capable of hunting adult bison: wolves, grizzlies and Folsom or Plano point-wielding men. Wolves of one sort or another had always been there to hunt weak or vulnerable bison, so their presence does not explain this behavioural shift, nor does that of grizzlies because carnivorous bears had also long been present in North America. Humans are thus the only force capable of 'making' the plains bison.

There is clear evidence in the fossil record that the Plano people were herd-driving by about 12,000 years ago. In order to hunt buffalo this way you must have beasts that associate in herds. It may be that the earlier Folsom hunters made the lives of individual buffalo, or those living in small groups, so precarious that by 12,000 years ago bison had already begun to congregate in larger numbers.

Archaeological studies from the mountainous country in the west, such as the Bighorn Mountains of Wyoming, show that hunting pressure may indeed have been intense from an early date. They reveal that long-horned bison once lived there, but that they apparently disappeared along with the rest of the megafauna about 13,000 years ago. Bison can clearly survive in such country as their modern presence in Yellowstone illustrates, so their absence from such areas in earlier Indian times is puzzling. It seems possible that the rugged terrain prevented them from forming the great herds that constituted their best defence against the human predator, and that the smaller groups were simply picked off one by one by Indian hunters.[8]

Stephen Pyne, that great historian of fire, thinks that an even more curious synergy existed between bison and Indians. Bison needed grassy plains in order to benefit from their new herding strategy, and it seems that such plains increased through the Indian use of fire over the past thousand years. It was about then that the bison crossed the Mississippi, for example, and by the sixteenth century fire had opened enough forest for them to invade the south. By the seventeenth century the same thing had happened in Pennsylvania and Massachusetts and the bison penetrated there too. An alternative argument is that reduced hunting pressure resulting from a disease-devastated post-Columbian Indian population permitted the bison expansion. All this leads me to believe that while a symbol of the 'wild west', beloved of the wildest of 'wild' Indians and a victim of the likes of Buffalo Bill, the bison is a human artefact, for it was shaped by Indians and its distribution determined by them.[9]

Studies of fossils reveal something of the timing of the size reduction of bison as well as changes in the shape of their horns. Judging from the diameter of their eye sockets, paleontologists have determined that bison have been shrinking in size throughout the last 12,000 years, shrinking most over the last 5000 years. The study of their horns reveals a somewhat different picture, for these have steadily reduced in size over the past 12,000 years. Similar dwarfing is also seen in many other large mammals including bighorn sheep, elk, moose, musk ox, bears, American antelope, wolves and wolverines to name a few. All decreased in size over the last 10,000 years.[10]

The size that a creature reaches is very much dependent upon its nutrition. Specially fed bighorn sheep and elk can attain enormous dimensions and have superb horns and antlers. Furthermore, all species have fluctuated in their average dimensions through time, and the late Pleistocene—between 100,000 and 10,000 years ago—was a time of true giants. Dale Guthrie calls the great herbivores of this golden age megabucks, megabulls and megarams.[11] Because of the influence that climate and food availability have on size, these two factors were thought to be the sole cause of the dwarfing of North America's mammals. I think, however, that hunting by or competition from humans exerted the greatest influence and precipitated much of the dwarfing experienced by large mammals over the last 12,000 years.

Human hunting can clearly lead to size reduction, as has happened in many historic fisheries, for neither cod nor lobster are what they were in grand-dad's day. Hunting pressure can cause dwarfing when it is intense and focused on the largest individuals, resulting in selection for early-maturing dwarfs. These are animals that cease to grow early in order to reproduce before they are taken by a hunter. Just as with cod, human hunting, I think, acted preferentially against the largest individuals of all Guthrie's 'mega' species. The size change among large ungulates may have occurred because most hunters tended to select the largest individuals, just as modern hunters do today. In smaller species such as armadillos, however, the dwarfing occurred because the larger individuals found it harder to hide from humans.

In both Australia and North America dwarfing of species as a survival response coincides with the arrival of humans and not with climate change. In Australia, kangaroos and possibly other large marsupials had begun to shrink by 40,000 years ago while in the Americas dwarfing began 12,000 or 13,000 years ago.[12] This is still an active area of research, and, with two very different theories to test, advances can be expected in the near future.

EIGHTEEN

THE RISE OF CULTURES

We must now turn to the people that colonised the continent. Not all native peoples of North America are descendants of Clovis. At least two other migrant groups established themselves on the continent before 1492, the ancestors of the Nadene and the Aleut–Inuit cultural complexes. The Amerindian peoples, as descendants of Clovis and Nadene are known, are often designated as 'Indians', yet they are quite distinct genetically and linguistically. Nadene people look more like modern north Asians than the Amerindians, suggesting that they branched off from their Asian ancestors more recently. Studies of the distinctive Nadene language family, and of genes and archaeology, indicate that they entered the continent around 9000 years ago. They must have come across the sea from Asia, as Beringia was submerged by this time.[1]

The traditional Nadene strongholds were the boreal forests and western coast of Canada where Nadene groups such as the Athabaskans and Haida were based. A few hundred years before the arrival of

Columbus, one Nadene group adapted to life on the Great Plains. Moving south, they migrated as far as the desert country of the American southwest, on the way becoming Navajo and Apache. The flat 'Asian' faces that gaze proudly out at us from the portraits of war leaders such as Geronimo reveal the Nadene origin of these groups. They were a people comfortable with climatic extremes, able to make a living in hostile regions such as the boreal forests, Great Plains and deserts that were too harsh for others to thrive in.[2] Either these people or the Inuit brought the bow and arrow with them. There is evidence that this weapon was present in North America around 4500 to 7000 years ago. Its absence in the earlier American archaeological record is remarkable given that it has been present in Eurasia for 20,000 years.

The Aleuts and Inuit represent a more recent development, for linguistic, genetic and archaeological evidence indicate that they arose around 5000 years ago. These peoples possess a greater number of north Asian physical characteristics than the Nadene (such as the Asian eye fold, blood group similarities, and a very flat face), as well as sharing some striking cultural similarities. The Aleuts, for example, who inhabit the Aleutian Island chain off Alaska, practise a sophisticated system of acupuncture very similar to that of the Chinese and which bears the same rationale; in both cases the primary reason for treatment being to let out 'bad air'. As with the Chinese, massage was also an important aspect of medical care among Aleuts and, not surprising in the case of the hunting Aleut, both peoples had a detailed knowledge of anatomy.[3]

Examination of Aleut–Inuit stone tools reveals that the cultures may have originated in Siberia. Some Inuit made their homes in some of the harshest habitats occupied by humanity; where winter lasts eight to nine months, rainfall and productivity are perilously low, and where just fifty days per year are frost-free. Regardless of their precise point of origin, in an important sense the Inuit are neither a North American nor Eurasian people, for like the polar bear and musk ox they have a true circumpolar distribution. As far as we can tell, before around 3000 years ago the Alaskan Inuit did not hunt seal, make skin boats or use dogs to haul their sleds, yet despite these limitations and the difficulty of their environment, they spread from Alaska to Greenland within about three centuries.[4]

The Inuit differ from all other pre-Columbian people in that they never lost contact, at least for extended periods, with their relatives in Eurasia. Inuit archaeological sites are full of Asian material culture. The artefacts indicate that the cultural powerhouse of the American Arctic lay to the west in Eurasia, for Eurasian archaeological sites show scant evidence of importations from the New World.

Other important Eurasian innovations to travel with Inuit to the New World include pottery making, which reached Alaska from Siberia by 2700 years ago (long after it had been independently invented in Central and South America), and iron tools, which had arrived from Siberia by the time of Christ. Over the last two millennia a succession of Asian innovations from slat armour to daggers has followed the same route into Alaska. The Inuit even received goods from the people of Europe. Archaeological excavations on Ellesmere Island have yielded pieces of chain mail, cloth and iron rivets, presumably acquired from the Viking settlements of Greenland. Thus not even transatlantic contacts were a new thing to these globe-circling peoples. Were we to see American history through their eyes, the arrival of Columbus in 1492 may well have been a rather unexceptional event.[5]

While Inuit cultural contacts were extensive both east and west, they enjoyed no such relations with the people to their south, for contacts between Indian and Eskimo have always been minimal; active avoidance alternating with hostility. The very name 'Eskimo' supposedly derives from a derogatory Indian term connoting 'eater of raw meat'. This hostility proved to be a barrier to the transmission of goods and ideas, with even such a desirable acquisition as the bow and arrow faltering at this breach, for it seems that the Inuit and Indians obtained this most important weapon at different times.[6]

One other important point to make about this history of immigration from Eurasia is that each successive wave of human invaders found their niche in a more marginal part of North America. The Amerindians were first in and inherited the best of the continent. The Nadene followed, and to them fell the difficult northern forests, some of the north-west coast and in time portions of the deserts and plains. Last came the Inuit, who despite their sophisticated technology and links with Eurasia could find a home only in the Arctic wastes. This is, of course, a very different pattern of human settlement from that arising after 1492. It may be more consistent with the patterns created by other mammal species as they crossed the Beringia land bridge over millions of years, although a detailed analysis of the zoogeography of the fossils is yet to be done.

One other strong parallel of pre-Columbian humans with earlier invasions exists in the fact that *no* human invasion ever went the other way—from North America to Europe. Here is a pattern of migration that persisted for 33 million years and would persist into the future. The Nadene expansion in the plains and deserts, however, echoes the fate of migrant species from smaller or poorer lands for, like sloths and camels, they found a home in some of the most challenging environments the continent had to offer.

By 8000 years ago a marked diversification of Indian cultures began, when groups living in various parts of North America started developing technologies and lifestyles to suit their particular location. Early evidence for regional specialisation is often difficult to discern in the archaeological record, for there are few old sites and nearly all were inhabited by people of the inland. The coastal sites are almost all underwater because the sea has risen more than 100 metres over the last 15,000 years. One coastal site in Labrador, however, was preserved by a fluke of geology. It too would have been drowned except for the fact that Labrador has been rising ever since the ice that covered it began melting 15,000 years ago, meaning the coast is now over 100 metres higher than it was then. With the rising land and sea pacing each other, parts of Labrador that were coastal 15,000 years ago are still on the coast today.

The Labrador site includes a grave that was built about 7500 years ago. It was filled with the tools of a seafaring people, including a spearhead made of antler for a toggling harpoon. The toggling harpoon is a specialised weapon with a detachable foreshaft and a point that swivels on impact to embed itself firmly in the prey. It is surprising to find such a sophisticated weapon used for hunting marine mammals at such an early date in North America. It suggests that as yet we know little about the early phases of Indian adaptation to life in the New World.

At about the time Labradoreans were making their toggling harpoons, Americans elsewhere were creating different kinds of weapons and some at least were pointed at each other, for at this time we find the first definitive evidence of warfare in North America. Three skeletons from Mulberry Creek, Tennessee, with stone points embedded in their backbones or lying among their ribs, constitute our earliest proof. Perhaps the rise of increasingly different cultures and lifestyles, along with resource shortages following the demise of the megafauna, led to conflict.[7]

Increasing cultural differentiation and a rise in population seem to have occurred in the eastern part of the continent during the altithermal period, between about 7000 and 4000 years ago. At this time temperatures in the middle latitudes of the northern hemisphere averaged about a degree higher than at present and rainfall decreased in the west of North America. The altithermal had varying effects on Indian cultures across the continent. Some populations, particularly those in the east, benefited from the change, but those living in the American south-west suffered and it seems probable that the south-west was depopulated during the early altithermal, around 7000 to 5500 years ago.[8]

It is symptomatic of North America's amazing capacity to amplify climatic change that this event should have such a profound effect on the Indians. As a global climatic event the altithermal was somewhat trivial, going largely unnoticed in the southern hemisphere and other regions. In North America, though, it dictated the fates of entire peoples, destroying desert cultures while assisting those further north.

By about 4500 years ago some Indian groups seem to have settled down in permanent camps and specialised in gathering and processing new kinds of food. Evidence of trade and more stratified societies becomes more apparent at this time. These changes mark the end of the archaic phase of Indian cultural evolution and from here on we see an intensification of trends towards cultural diversity, specialised food gathering and production and more complex societies.

By the time of Columbus the people of North America had created an astonishing variety of human societies. All had been shaped by the opportunities and the imperious demands of the North American continent as mediated through human ingenuity, and almost all had developed in isolation from the inhabitants of other lands. Taken as a whole, they represent a constellation of human possibilities—a corpus of cultural experiments—that really does represent a New World. In their encounters with such people the European explorers came as close as anyone ever will to meeting the alien societies so vividly brought to life in science-fiction fantasies such as *Star Wars*. Today the sheer difference and ingenuity of many American cultures can only be glimpsed at second-hand, through the writings of explorers such as John Smith, who around 1608 met some Virginian Indians:

They adorn themselves most with copper beads and paintings. Their women, some have their legs, hands, breasts, and face cunningly embroidered with divers works, as beasts, serpents, artificially wrought into their flesh with black spots. In each ear commonly they have three great holes, wherat they hang chains, bracelets, or copper. Some of their men wear in those holes a small green and yellow coloured snake, near half a yard in length, which, crawling and lapping herself about his neck, oftentimes familiarly would kiss his lips. Others wear a dead rat tied by the tail. Some on their heads wear the wing of a bird, or some large feather with a rattle. Those rattles are somewhat like the shape of a rapier but less, which they take from the tail of a snake. Many have the whole skin of a hawk or some strange fowl stuffed with the wings abroad. Others a broad piece of copper, and some the hand of their enemy dried. Their heads and shoulders are painted red with the root *pocone* brayed to powder, mixed with oil; this they hold, in summer, to preserve them from the heat, and, in winter, from the cold. Many other forms of paintings they use, but he is the most gallant that is the most monstrous to behold.[9]

The foremost question that arises in the mind of the historical ecologist confronted with such a phenomenon is why Indian cultures diversified when and where they did. One way of investigating this is to look for similar patterns (albeit biological rather than cultural), among other recent arrivals. This then raises another question: does the North American continent promote rapid diversification in its recently arrived immigrant species?

When I first visited North America in 1983 I was keen to see as many of its native mammals as possible. I bought a field guide that stayed by my side wherever I travelled and which I read at night for pleasure. While so occupied one evening I absorbed the startling information that seventy-four 'species' of grizzly bear had been described by early

American biologists, and that many of these bore curious and long-disused names. I was intrigued by the so-called *Ursus hoots*, yet even today I am unsure how this bear came by its outlandish name. I did find, however, that *Ursus hoots* and many of its *confrères* were the work of Dr C. Hart Merriam of the US Biological Survey.[10] Merriam was not a skilled taxonomist and many of his 'species' were the result of poor science, for he often mistook individual differences for differences between species, which resulted in what taxonomists call 'oversplitting'. Yet contemporary checklists of North American mammals informed me that up to a dozen of Merriam's names might represent discrete geographic populations.

Dubious about the validity of even this lesser claim, I searched museum collections at every opportunity in order to examine grizzly skulls. To my astonishment I found that considerable variability did indeed exist. The skulls revealed that, despite the fact that grizzlies had inhabited North America south of Alaska for just 13,000 years, up to a dozen genetically discrete populations of grizzly had begun to form in North America by 1492. In terms of their ecology grizzlies are a bit like hunter-gathering humans, for both are large omnivores with considerable energy demands. It takes a lot of land to feed such creatures and that land must be rich in resources. The diversity of distinctly different grizzly populations—and the richness of North American ecosystems it implied—astounded me.

How, I wondered, did such large mammals diversify so quickly? The answer lies not only in the richness of the continent but in the 'ecological release' experienced by the bears in their new homes. With few competitors and a huge variety of resources they quickly diversified and adapted to local conditions. Indians of course do not represent incipient new subspecies as grizzlies do, but for millennia before 1492 they were adapting through cultural change to local conditions even more rapidly than grizzlies were through natural selection. As a result both

Indians and grizzlies have developed exceptional diversity in a very short period of time.

Whenever a species arrives in a new habitat, a series of evolutionary forces come into play that have a dramatic effect upon it. The nature of a founding population has a considerable effect upon the process. Such populations are never truly representative of the populations from which they are drawn, and this leads to a founder effect. It may have been, for example, that there were no expert bow and arrow makers among the first people to cross into North America and this craft may have been lost to the New World through this founder effect.

If the new homeland that a species invades is essentially an open field, with few competitors but lots of food, then the species goes through a period of ecological release. For creatures such as birds this can result in a population that is more varied (in beak shape, leg length or size) than the parent population. This happens because in the absence of competition almost all variants can make a living of some sort. This phase of the evolutionary process is typically brief, occurring over a few decades or centuries. Then, as the open field is filled, individuals in the variable population begin to be selected for various traits and begin to adapt to local conditions. Long beaks and long legs may be favoured in one environment and short beaks and short legs, or some combination of both, in another.

Thus begins the long and final phase of the adaptive process that makes species. Known as evolution by natural selection, the process can be thought of as a great centrifuge, throwing apart the geographically separated portions of a once similar people or species, creating diversity out of uniformity. The richer and more diverse the environment is, the faster the centrifuge can be thought of as turning. The centrifugal force is felt most strongly after the closing of the frontier, and for organisms like large mammals its results via natural selection are evident only after thousands or tens of thousands of years. For humans, however, who adapt through learned cultures, the process can happen swiftly.

This process wrought varying results among Indians inhabiting different parts of America. In some regions little changed since Plano times, for Indians living on the Great Plains continued to drive-hunt buffalo right down to the historic period, having perfected the technique 12,000 years earlier. Others, such as the nomadic 'Digger Indians' of the Californian deserts, continued to live in family-sized bands and to subsist by hunting and gathering small items. Elsewhere in America people developed a tradition of hunting smaller game but placed little reliance on vegetable food. Various rock shelters in Alabama that were occupied between about 10,000 and 8000 years ago provide evidence of this, for they yield many hunting tools along with the bones of deer, turtles, turkeys and squirrel, but no tools used to prepare vegetables. Hunting remained important to most Indians right down to the historic period, partly because North American Indians had only two domestic animals, the dog and turkey, which supplied relatively little protein.[11]

It is almost undisputed that the gathering of vegetable foods by Indians increased in importance over time, most likely as a result of an increasing scarcity of animal food. Nuts and acorns were among the first vegetable foods to be harvested *en masse*—beginning about 8000 years ago there is evidence in the archaeological record of an increasing reliance upon them. This is hardly surprising, for such nuts had been selected over millions of years for their nutritional value and accessibility by squirrels. They were, in effect, an already domesticated crop, but it was squirrels rather than humans that had done the selecting.[12]

Had we been able to visit the coast of California between 5000 and 400 years ago we would have seen a remarkable sight. We could have wandered into large, permanent villages, some perhaps consisting of a thousand or more people. There we would have found a ruling elite, a working class, ritual specialists and skilled craftsmen and women, as well as extensive evidence of trade. While this kind of society may seem familiar, the thing that made the Californians special was that nowhere around these towns would you have seen fields or pasture. All of this social complexity was generated in the absence of agriculture.

Eurocentric scientists long believed that complex civilisations are the exclusive reserve of an agricultural people, that elites and specialist craftsmen can only be supported out of an agricultural surplus. Now scientists are beginning to ask how it is that the Californian Indians, along with a few other North American groups, came to be the societies that broke the rules. They certainly broke them in a big way, for in pre-Columbian times California supported about 10 per cent of the total population of North America north of Mexico.[13]

Unfortunately, Californian Indian cultures were disrupted from a very early date by European intervention. First Spanish then American colonists assailed them. The final blows were delivered during the gold rush. Between about 1850 and 1870 California saw the worst massacres of indigenous peoples ever to take place in the United States. Tens of thousands of Indians perished at the hands of American frontiersmen, miners and camp followers. Thus researchers interested in these intriguing cultures have only a few early accounts and the archaeological record to guide them. These reveal that many ancient Californians gained much of their nutrition from a bountiful sea containing marine mammals, fish and various invertebrates such as shellfish, which were harvested from plank canoes that also greatly facilitated coastal trade.[14]

The development of maritime technology in these west-coast societies, incidentally, was not without environmental cost, for two North American extinctions occurred within the last few thousand years before 1492, both involving marine species living on the west coast. Around 5000 years ago a flightless, ocean-going duck of the genus *Chendytes* vanished from coastal California. It presumably became extinct when Californian Indians developed watercraft that allowed them to reach the offshore islands where it bred. Likewise, the eight-metre-long Steller's sea cow (a relative of the dugong) disappeared from the west coast of America about 2000 years ago, although it survived around the Bering Islands until discovered by Russians in the eighteenth century. As with the flightless duck, the development of watercraft such as plank canoes may have led to this huge beast's demise.[15]

Although they were not agricultural societies, plants played a vital role in Californian Indian life, and no resource was more important than the acorn. Acorns are not an easy resource to utilise, requiring crushing, leaching and boiling before they can be consumed by humans; but the resulting meal has the virtue that it can be stored for considerable periods. While lower in protein and carbohydrates than agricultural crops, acorns are very high in oil. Maize and wheat contain about 1.5 per cent oil whereas acorns hold 20 per cent in their raw state and 10 per cent as cooked meal.[16]

Somewhat similar large-scale hunter-gatherer societies were found further north on the west coast, and these survived rather better than the Californians. In the far north-west the Nadene Haida and related groups spent their lives shifting between two or three permanent villages composed of plank-built long houses. Whenever the tribe moved they left the structural supports of their dwellings in place but took the plank cladding in canoes to the next village site. Although not as large as the prehistoric towns of California, their historic villages could accommodate several hundred people. The mainstay of their economy was fish, such as salmon and herring, which were harvested in a frenzy of activity when they spawned, then dried and preserved to be eaten for the rest of the year. Marine mammals such as whales, which were hunted from large, ocean-going canoes, were also important and it is likely that the fat of these creatures along with fish oil provided a vital dietary supplement.[17]

Because the Haida survived into modern times, we know rather more about them than we do of the Californians. They had strict hierarchical societies with three levels or classes—chiefs and their families, commoners and slaves—with slaves representing as much as a third of the population. These societies were unusual, however, in that not only did every individual know to which class they belonged, but each their individual rank. In a village of 500 people someone would know that they were number 324 in rank, and be able to tell you the ranks of those above and below them.[18]

The archaeological record provides evidence that other prehistoric North American hunter-gatherer societies developed greater social complexity, size and splendour than even the Californians. The greatest of these were the Adena and Hopewell peoples, early mound-builders in the Mississippi and Ohio valleys, who developed the most impressive if mysterious of the North American Indian cultures.

The massive earthen mounds that form the material 'signature' of these two peoples have intrigued Europeans ever since the earliest crossing of the Appalachians. Thomas Jefferson was fascinated by them, and the rumour soon grew that they had been built by people who had come from, or had extensive contacts with, the Mediterranean region. How else, people asked, could such magnificent edifices have been erected in this wild continent? The truth was in fact far stranger, for the builders of the earthen mounds were hunter-gatherers.

It is no coincidence that the mound-building cultures sprang up along great rivers; waterways are conduits of nutrients that support abundant life. Fish and shellfish abound while the fertile floodplain soils give rise to a riot of plant life. Typically, such areas support population densities ten to a hundred times greater than the surrounding countryside. The Adena first began to erect mounds about 2500 years ago near what is now the town of Chillicothe, Ohio. Their mounds served several purposes, with one of their principal functions as a place of interment for their leaders. By around the time of Christ the Hopewell were also building mounds, and some of their mound complexes were prodigious.

Near Chillicothe, for example, thirty-eight mounds were erected within a rectangular enclosure of forty-five hectares. The number of man-hours involved in such erections is beyond the comprehension of most hunter-gatherers.[19]

Hopewell trade was extensive. Obsidian came from as far afield as present-day Yellowstone National Park, while shells and animal teeth were brought from Florida and copper from Lake Superior.[20] Hopewell cultural influence spread out along the trade routes and was soon felt as far afield as Wisconsin and Florida. The picture one gets of Hopewell culture is that of a powerful and sophisticated centre whose influence was felt throughout most of the continent but, despite its power, by AD 400 the Hopewell people had erected their last mound and traded their last seashell. Their culture was, for reasons still not understood, in collapse.

NINETEEN

THE TAMING OF TEOSINTE

About 1300 years ago the Mississippian people began to build again. These people, however, were different from earlier mound-builders. They were agriculturalists, and they did not use mounds as places of burial. Instead, the cultural elite of the Mississippian society lived atop the great erections. In various regions these people attained the population and societal complexity requisite to qualify as chiefdoms. Indeed one group even seems to have approximated a state. States differ from chiefdoms in that relations are no longer conceptualised as kin-based, but are divided into classes comprising royalty, nobles, commoners and slaves. Agriculture may well be a precondition for the transition from chiefdom to state, for no hunter-gatherer-based states are known.

About the time that William the Conqueror was sailing across the English Channel the people of Cahokia, just east of present-day St Louis in Illinois, were raising the largest earth mounds ever built in North America. Monk's Mound contains some 600,000 cubic metres of earth and rises to a height of thirty metres above the surrounding plain. About

120 subsidiary mounds were constructed around it, the entire complex covering thirteen square kilometres. The workforce needed to undertake this venture was huge and it is estimated that about 30,000 people lived in or around Cahokia. Not surprisingly, there is evidence of a complex hierarchy consisting of at least three levels—probably a paramount chief and his immediate followers, a lower tier of chiefs and the commoners. By between AD 1050 and 1250, when Cahokia was at its height, it was either a very complex chiefdom or a nascent state— the first to exist in what is now the territory of the US. Its size reflected this, for Cahokia was then as large or larger than any city in Europe.[1]

Despite the great size and complexity of societies such as Cahokia, the real centre of North American cultural evolution lay far to the south, in what is now Mexico. This area has always had large populations, and innovations made there have, since time immemorial, found their way north. Agriculture is a prime example, for it began in North America long before and far away from Cahokia, somewhere in Mexico. More, perhaps, has been written about the origin of agriculture than any other topic in New World prehistory, yet some profound mysteries still lurk at its heart. One of the most extraordinary concerns the early domestication of plants in the New World, specifically the domestic gourd, which was used to make containers, rattles and such like. Archaeological evidence, which was long widely accepted, suggests that gourds were being grown in Mexico 9000 years ago and in Peru 8000 years ago. Recent redating of the plant remains, however, suggests that actual domestication of the gourd began several thousand years later.

This early domestication might seem unexceptional were it not for the astonishing fact that the domestic gourd has no wild relatives in the New World—it originated in Africa. How then did the ancestral gourd make its transatlantic crossing? There is no evidence that African farmers were cultivating domestic gourds 9000 years ago, or that humans could cross the Atlantic at that time, so transmission by human means is unlikely. Indeed it would be preposterous to suggest that transatlantic voyages were occurring so long ago based on the gourd's presence in the Americas. Gourds, however, can float, and one can imagine a gourd drifting across the Atlantic from Africa to the Americas. Yet this hardly solves our problem, for its seeds could not grow where they washed ashore, as gourds need good riverside soils to put down roots.[2]

We are left with the rather feeble hypothesis that, one day in the early Holocene, some far-sighted Indian beachcomber found, among beach-wrack on the Atlantic coast, an African gourd that contained viable seeds. She or he then carried it inland and planted it in rich riverine soil, where it gave rise to all the gourds of the New World. Even more extra-ordinary is the fact that it would be thousands of years before anyone in the Old World, where gourds abounded, would do the same thing.

The earliest of all American plants to be domesticated was the squash. As with the gourd, dating has proved controversial. Archaeological evidence indicates that squash was being cultivated in Mexico by about 10,000 years ago, though new radiocarbon dating on actual remains suggests that those dates are a few thousand years too early. It is also claimed that a related species may have been cultivated in Amazonia at this time, while maize may have been cultivated in central Panama by 7700, or around 6000, years ago, depending on which set of dates you believe. The important point for our purposes here, however, is that all of these domestication events occurred south of the Rio Grande, where by AD 1500 around 25 million out of the continent's 30-odd million inhabitants lived. The early possession of agriculture would give these societies a lead in both technology and population that would persist well into the modern period.[3]

About 4500 years ago agriculture began to be practised in what is now the US. Initially, the plants grown there were an unlikely lot—sumpweed, goosefoot, knotweed and maygrass—along with the more familiar agricultural products of sunflower, pumpkins and gourds, the latter two having arrived as already domesticated plants from Mexico. Early evidence for the domestication of the then important crop sumpweed (it's now considered a pest) comes from an unlikely find in Kentucky—human turds preserved in Salts Cave in the quaintly named Newt Kash Hollow. As Percy Bysshe Shelley knew when he wrote 'Ozymandias', time is a great leveller. One wonders how the prehistoric turd's owner would have felt if he or she known that it would be their sole lasting contribution to the sum knowledge of humanity.

The deciduous forests of the north-east represent the most distinctive and full response to the extreme bimodal climate of North America. Their Indian cultures are therefore of particular interest for they should show adaptations to the peculiar constraint that these environments place on human societies. Unfortunately, the Narragansetts, Algonquians, the tribes of the six-nations league and others were hit hard and early by European impact and little is known about their agriculture and lifestyle. One fact that does shine through, however, is that many of these people had a distinct seasonal round different from anything seen in Eurasia. With the exception of some northern Algonquians in Maine all practised agriculture, planting their fields with maize, beans, squash, pumpkins and tobacco (each being southern North American in origin). While the fields were tended they inhabited large permanent villages of up to 1000 people that were usually surrounded with a defensive palisade, but in October they left these settlements, dispersing into small family groups to hunt on family-owned hunting grounds. They would meet a few months later to overwinter together, often at a different site from the agricultural village so that a good supply of wood could be obtained nearby.[4]

Many American Indian societies seem to have practised a similar mixed hunting–farming lifestyle, but such truly mixed economies are sometimes difficult to recognise in the archaeological record. Historic documents, for example, show that the Cocopa people of north-western

Mexico, who were primarily hunter-gatherers, opportunistically sowed seeds of two species of wild grass on the floodplain of the Colorado River to supplement their diet. Such practices are almost invisible in archaeology; nevertheless we must count this as one of the truly important adaptations of indigenous Americans to their unique environment, for it reflected and permitted full utilisation of their strongly bimodal world.[5]

Despite the early development of agriculture in Mexico it was not until around 3000 years ago that agriculturally based chiefdoms developed there. The earliest were the Olmecs, an enigmatic people of the region now known as Veracruz. The Olmecs left significant statuary littering the landscape, including 2.5-metre-tall stone heads made of basalt that had been transported over 100 kilometres. They also constructed large buildings for ceremonial use. Despite these achievements, Olmec influence was waning by about the time of Christ—its rise and fall spans less than 1000 years.

A welter of agriculturally based chiefdoms and states rose and fell in southern Mexico following the Olmecs, and the one thing they share is that all were short-lived. By about 1500 years ago the Mayan settlement of Peten had achieved the complexity of a state. It was followed by several other centres of similar complexity, yet by 1100 years ago most Maya city-states were in collapse, although the very last Maya settlement did not fully dissolve until taken by the Spanish in 1697. Most of the region has remained sparsely populated ever since.[6]

From about 3000 years ago, the town of San Jose Mogote in the Valley of Oaxaca was the centre of a minor chiefdom. Around 200 years later it had been eclipsed by the settlement of Monte Alban, which was larger and more complex. By about the time of Christ, Monte Alban had assumed the size and complexity of a state and it survived until about 1000 years ago, when it collapsed without replacement.

The Mexico Basin was home to the greatest state to arise in North

America—the Aztec empire. From a geological perspective the basin is a remarkable structure. At 2400 metres elevation, it was originally occupied by a large shallow lake. It appears to have been largely uninhabited except for passing hunter-gatherers until about 3000 years ago when settlers from Morelos moved into the area. As late as 2500 years ago the population of the basin numbered just 25,000, but just 200 years later it had grown to about 75,000. By this time social complexity commensurate with that of a well-advanced chiefdom had been established. Around the time of Christ's birth, the basin's population exceeded 100,000.

By about 1500 years ago the city of Teotihuacan in the Mexico Basin had evolved some of the complexity of a contemporary city. It may have housed as many as 200,000 people who lived in substantial stone-walled residential compounds arranged in the familiar contemporary American grid pattern. A main street ran through the town centre and the richer residents clustered along this prehistoric Fifth Avenue. Teotihuacan even had neighbourhoods occupied by guilds, with potters clustering in one area and obsidian workers in another. But what made the town seem even more 'modern' was the presence of ghettos, for in the western part of the city lived a community from Oaxaca whose members still made pottery typical of their homeland and who buried their dead in Oaxacan manner. Here, surely, was an embryonic New York City.

There is something else about Teotihuacan that is uncannily reminiscent of contemporary North America, for planned centres such as villages and hamlets were 'planted out' over the valley, all laid out in imitation of the centre. The standardised architecture and design of these prehistoric settlements bears an eerie echo of the towns of the twentieth-century American expansion, whose cities were built on the same grid with standardised street names, each state capital nestling a domed state house in its bosom. Around 1400 years ago Teotihuacan began to decline and the city was finally burned and destroyed 1250 years ago. This grand premonition of the urbanised continent lasted just 250 years. Climatic change, destructive agriculture and the rise of rivals such as Cholula have all been cited as the cause.[7]

The Toltec capital of Tula is situated on what is now an arid plain in the north of the Mexico Basin. Tula grew rapidly between 1250 and 1050 years ago, absorbing immigrants from both north and south, resulting in a peak population of 60,000, yet by 1000 years ago this city-state was also in precipitous decline.

There is one last society that must be examined in this brief survey of Mesoamerican states. The Aztecs, it seems, were bad neighbours from the very start. So bad in fact, that if we are to believe the words of conquistadors such as Cortez, many of their Indian subjects were delighted to exchange them for Spanish masters.

The Aztecs first arrived in the Mexico Basin around 800 years ago, having left their home in western Mexico for reasons that remain unclear. At that time there was little spare land in the basin and the Aztecs squatted on one piece of tribal land after another. Sooner or later, however, they were always driven off by their neighbours. Finally the people of Colhuacan permitted them to settle on their land as serfs, and all went well until a Colhuacan prince offered his daughter's hand in marriage to the chief of the Aztecs. He arrived at the wedding only to find an Aztec priest dressed in his daughter's flayed skin. The Aztecs had decided she was more useful as a sacrifice than a bride. Needless to say the Aztecs once again resumed their wanderings, but finally found some unoccupied swampy islands on the western shore of Lake Texcoco. There, around AD 1330, the twin Aztec towns of Tenochtitlán and Tlatelolco were founded. By 1428, by virtue of political intrigues and alliances, the Aztecs and their allies the states of Texcoco and Tlacopan had established themselves as the rulers of the basin. By 1500 the alliance controlled a territory of

324,000 square kilometres and a population of ten million people, and by 1520 the Aztecs had subjugated their erstwhile allies and effectively ruled their empire alone.

At its height in 1520, Tenochtitlán covered about thirteen square kilometres and was home to about 200,000 people, with a population of perhaps a million living in the Mexico Basin. When the Spanish first saw the city it was five times the size of London. So splendid was it that it astonished the first Europeans to observe it.

On his first view of the place the famous conquistador Bernal Diaz del Castillo wrote: 'With such wonderful sights to gaze on we did not know what to say, or if this was real that we saw before our eyes. On the land side there were great cities and on the lake many more. The lake was crowded with canoes…' Upon entering the city Diaz saw a zoo that exceeded any menagerie in Europe in its magnificence and the diversity of its inhabitants, palaces to rival the Alhambra and markets that outshone those of Venice. 'Some of our soldiers had been in many parts of the world, in Constantinople, in Rome and all over Italy,' Diaz continued, though none had seen anything like the magnificence of Mexico.[8]

Despite their greatness, or perhaps because of it, the Aztecs subscribed to a fatalistic, even paranoid view of the world. They believed that the universe had already been utterly destroyed by vengeful gods on four earlier occasions. If the terrible Aztec gods were to be prevented from completing a fifth destruction they had to be supplied with a steady diet of human hearts and blood. At the dedication of their renovated Great Temple in 1487, bloodspattered Aztec priests worked ceaselessly for four days ripping the hearts out of 14,000 people.[9]

Aztec beliefs may seem savage and bizarre to us, but there might be a very good and distinctively North American reason for such practices. Just a brief survey of the city-states of southern North America suggests that their gods were indeed vengeful. The Olmecs, Oaxacans, Maya and

Toltec had all experienced meteoric rises before the Aztecs and then just as dramatically perished. The remains of their great cities littered the landscape inherited by the Aztecs, who would also have heard through traders of the great fallen cities of the Maya. Perhaps they saw their fate most clearly in the ruined Toltec capital of Tula, which they looked on as the remnant of a golden age. Aztec legend has it that Topiltzin, the last ruler of Tula, was something of an intellectual pacifist. He abhorred human sacrifice and his reluctance to feed the gods enraged them, so they destroyed Topiltzin and his city. The Aztecs so admired Tula that they dismantled its monuments in order to adorn their own cities. More ominously, they associated Topiltzin with their own divine Quetzal-coatl.[10] The Aztec motto may well have been 'there but for the flow of blood go I', for there were to be no surviving North American cities, just great fallen citadels amid jungle or desert to remind one of the fate that eventually awaits us all.

I have long pondered just why it is that all of the American city-states and even chiefdoms seem to have been so fragile, and to have so often fallen without immediate replacement. Admittedly, many Eurasian cities have proven equally transient, but some such as Rome, Constantinople, or even London have been far more enduring. The rise and fall of North American societies may in part be related to the great climatic trumpet. Environmental conditions can change rapidly, driving people from the south-west as conditions warm, or allowing Inuit to take over from Indians in the north as it cools. As these changes in rainfall, temperature and plant growth play out across the continent they may open and close opportunities for various species, including humans. Agricultural societies may be particularly vulnerable, surviving only during the brief window when local conditions are right.

Other factors may also have been at work, for in centres such as the Mississippi Valley, the Mexico Basin and the south-east coast of Mexico, city-states and chiefdoms have continually risen and fallen. Shortages of food and building materials brought about by over-exploitation have doubtless played a role too, as perhaps has competition with other peoples.

This 13,000-year-long excursion through the prehistory of the New World has brought great insights to our ecological history, for the period is close enough to us that we have been able to examine evolution in action. We have been able to speculate about how predators affect their prey, how species change to fill an extinction vacuum and how the founder effect can influence future populations. In short, we have been able to examine, on a manageable scale, those forces that create ecosystems. For those interested in generalising about the past, however, it is disconcerting to realise just how different the extinction of the megafauna was from earlier extinction events.

As far as our species goes, we have seen how mammoth hunters transformed themselves into diverse cultures: Aztecs lording it over empires and vast city-states, Haida thriving on the maritime resources of the Pacific Northwest, Mississippi mound-builders who were both hunter-gatherer and agriculturalist, and the desert hunter-gatherers of the south-west. All four levels of human social complexity: simple hunter-gatherer bands, transitional societies, chiefdom and state, had evolved and flourished in North America, along with a breathtaking array of local economies.

Taken as a whole, the cultures spawned in the New World differed in several ways from those of Eurasia. A number of complex societies were based not upon agriculture, but upon hunting and gathering. Other groups used agriculture when it suited them then abandoned it when it did not in a way that is utterly alien to modern Europeans. Agriculturally based chiefdoms and states, furthermore, were slow to develop in North America and short-lived, often collapsing without replacement, the territories of entire states sometimes reverting to wilderness.

The end of pre-Columbian America would commence in the same

southern region that was its greatest strength. It would be heralded by the arrival of creatures that the Indians at first mistook for gods. In reality they were avengers for the environmental destructiveness of an earlier pre-Columbian age—the first horses to paw the ground of the New World since their kin died at the point of an Indian spear some 13,000 years earlier. Had the conquistadors been met by mounted Aztec battalions, the result of the war of the worlds might just have been different. We must now farewell this earlier America, for the date is 11 October 1492, and the morrow will dawn upon a very different age.

Act 5

1492–2000

•

In Which America Conquers the World

Twenty

ALTERNATIVE AMERICAS

This final act in the ecological history of North America spans a mere 508 years; just an eye blink—albeit an action-packed one—in the vastness of time. The period is characterised by two phenomena; one old, the other startlingly novel. The old concerns immigration from Eurasia. It's a pattern we have seen over and over again during this saga, but now it reaches a fever pitch. The new relates to the United States of America, which at the end of this period becomes the most powerful nation on the globe. Not since the Eocene has North American influence been felt so widely.

The written record allows us to see in much greater detail the environmental and evolutionary change induced by immigration and emigration. We can also examine what happens to a variety of immigrant human cultures as they respond to their New World. These cultures are just 500 years into the great experiment of becoming truly North American, but the degree of change they have undergone and the alteration they have brought to the face of the continent are astonishing.

Although the historical period provides more information than ever before, its examination is a perilous business. This is because, from an evolutionary perspective, reading the recent past is a bit like examining, from very close up, a newspaper photograph that has been enlarged a hundred times. We can see the individual dots clearly, but it's an effort to make out the bigger picture. The picture is still there—it's just that much is lost in the excess of detail. This problem occurs in part because in the geological 'present' we often don't see broad patterns of evolutionary adaptation but instead witness the struggle of the individual in a new and fast-changing environment. Without the benefit of time as a yardstick it is difficult to distinguish dead ends-to-be from enduring strategies. If that were not enough, human societies assume an overwhelming importance in this final act. Historical ecologists must be cautious when dealing with them, for human societies are not captives of their environment; they can change themselves by acts of will. And it's not just historical ecologists who must proceed with caution—much of our response in evaluating cultures is itself so culturally biased (patriotic or prejudiced) that it's very difficult to deal with them objectively. But go forward we must, as bravely and only a little less blindly, perhaps, than Columbus himself.

A 60-million-year tradition of immigration has brought almost everything—from elephants to Italians—to North America. The Columbian invasion of 1492 was part of that tradition and was therefore in accordance with the rules of zoogeography as played out over the millennia. The Columbian invasion is nevertheless exceptional in that it came from Europe, across the Atlantic on a route that had lain largely unused ever since the DeGeer land bridge foundered some 45 million years earlier. This single event makes it almost unique in the last 45 million years of Earth history, for virtually every other Eurasian invasion of the Americas had come from east Asia, across the Bering land bridge or the sea that at times covers it.

Asia's biological importance makes Columbus even more enigmatic, for ever since the *grande coupure* of 32 million years ago Eurasia (and particularly east Asia) has been the world's solitary 'biological superpower'. The region is so rich that although it comprises just one seventh of the globe's land surface it is home to over half of humanity. Quite apart from their sheer size and the biological resources available to them, historic east Asian societies have some characteristics that, on the surface of things, should help them to dominate, for they have centralised power structures enabling them to mobilise enormous national effort.

It was, after all, Asians who peopled the Americas 13,200 years ago, inheriting in one swoop 28 per cent of the available land surface of the globe. Beginning 6000 years ago another Asian people began moving into south-east Asia and the Pacific Islands. By 800 years ago they had reached as far as New Zealand, spawning the Polynesian and Maori cultures. The vast island realm occupied by them stretched across two thirds of the globe, from Madagascar in the Indian Ocean to Easter Island in the eastern Pacific. By the Middle Ages Asiatic people had extended even further, colonising a region extending from central Europe to Tierra del Fuego. In total, over 60 per cent of the Earth's habitable land surface was then occupied by people of Asian origin.

Despite the eastern victories of the likes of Alexander the Great and the Romans, the ecological history of Europe looks much less impressive when viewed over millennia, for Europe has long acted as a cul-de-sac and last refuge for various vanishing peoples. Neanderthals entered Europe from the east over 100,000 years ago, but were displaced by westward-migrating Cro-Magnons, until the last lonely clans perished on Gibraltar some 29,000 years ago. By 6000 years ago the Celts had arrived from the east, but they in their turn were pushed west by other, newly arrived groups, until they found a last refuge on Europe's westernmost isles.

By the fifteenth century the Europeans had been reduced to a rump by eastern invaders. The Iberian Peninsula and eastern and central Europe had fallen into Asian hands, while later, in 1683, the Turks besieged Vienna. It is edifying to remember that before he set sail for Virginia, explorer John Smith was fighting Tartars and Turks in what is now Germany and Hungary.

So why was Columbus not Japanese or Korean or Chinese? All of these nations are located closer to the New World than is Europe, and all have access to it. An Asian Columbus would barely have needed to sail out of sight of land before encountering the temperate and bountiful west coast of the North American continent. The journey was certainly possible, for during the nineteenth century straying Asian mariners at times reached as far south as California.

Various authorities have given views on this vexing subject, the latest being Jared Diamond in his book *Guns, Germs and Steel*. Diamond notes that while there are few east Asian states, there are many (much smaller) European states. If each state is assumed to have an even chance of making the discovery, then by virtue of their larger number a European state was likely to be the lucky one. While this argument makes sense, it begs the question of just *why* there are so many European and so few Asian states. This is an intriguing issue but to pursue it here would lead us into a discussion of the ecological history of Eurasia, which would entail writing another, quite different book.

An enormous windfall of resources fell into the hands of the Europeans upon the discovery of the New World. At first these were used to fight battles at home. Philip II of Spain, for example, used the first wealth from Mexico not to fight on the eastern boundary of Europe, but to foment European tribal wars with England. Nevertheless the tide would eventually turn, with the new-found wealth boosting trade and strengthening the merchant class, hastening the process of democratisation that was to reach full flower in the New World in 1776.

Europeans had been travelling to the New World since at least the tenth century, when Vikings attempted a short-lived settlement in Newfoundland, while Basques had probably visited the Newfoundland banks in their search for cod long before 1492. Despite these early contacts and voyages the Europeans were unable to gain a foothold to settle in the New World, though things were to change following Columbus's 1492 voyage to the Caribbean, for in his wake numerous European nations would establish their own colonies.[1]

Three nations, whose colonists arrived at different times and for different reasons, would play a significant role in this most recent peopling of North America. Each would experience a very different kind of frontier, and the nature of that frontier would be largely determined by the relationship that each established with the indigenous people it encountered. The Spanish led the way, developing colonies at a time when their own peninsular landmass was newly wrested from the Moors, and when their culture had just emerged from feudalism. Indeed, they carried the remnants of feudal Europe with them, and used them to build their New World empires. Around a century later the French and English followed, their settler societies emerging from a pre-Enlightenment Europe, carrying with them pre-Enlightenment attitudes and institutions to the New World.

These three streams of colonisation produced very different responses to their new-found home, each reflecting a unique interaction between culture and environment. They offer us a splendid opportunity to examine the interplay between ecology and culture in what is in effect three grand natural experiments. Those experiments reveal just how varied, in the short term, the responses of immigrant populations to their environment can be, and how quickly one fortuitous combination can come to dominate.

Before these great experiments commenced, the Europeans christened their new-found world. From our perspective Columbia would seem to be the obvious choice. Instead, the all but forgotten

Amerigo Vespucci, who published an account of a journey to the New World in which he claims to have beaten Columbus to the American mainland, was to have the honour. His account, long suspected as fraudulent, was read and believed by a group of scholars based at St Dié in the Vosges Mountains in eastern France, near Switzerland. Headed by Martin Waldseemüller, they were busy in their mountain retreat fashioning a new system of geography for the entire planet. In 1507 they published a huge map, nearly two and a half metres in length, along with a small globe. It was on this latter work that they printed the name AMERICA across the southernmost of the two newly discovered landmasses. Thus North America was christened by default. Many have protested the decision—Ralph Waldo Emerson called Vespucci 'the pickle dealer at Seville…who managed in this lying world to supplant Columbus and baptize half the earth with his own dishonest name'— and yet 'America' has stuck.[2]

Recent historical and even palaeontological research indicates that Vespucci may not have been such a great liar after all. The palaeontological discoveries concern Vespucci's observation that on 10 August 1503, on his supposed fourth voyage, he saw 'very big rats and lizards with two tails' on an island off Brazil. Although the two-tailed lizards remain a mystery, a few years ago fossils of large rats named by their discoverers *Noronhomys vespuccii* were found on the island of Fernando de Noronha off the coast of Brazil. The rats became extinct, it seems, soon after Vespucci's visit, perhaps because of black rats brought by the Spaniard.[3]

It is no accident that Spanish colonisation began in the Caribbean, for islands are ideal places for immigrant populations to gain a foothold. They often lack the diseases and pests found on continents, and immigrants to islands are able to subdue the relatively few, isolated indigenes. Islands also act as large, predator-free enclosures in which semi-wild livestock can proliferate free from competition and disease, providing an abundant resource for the new colonisers. In short, islands offer special opportunities for control in an arena where competition is limited—a valuable advantage for a society trying to establish itself.

The transplanted Spanish colonists flourished in their insular setting, but were even more fortunate when they penetrated the mainland. There they encountered highly organised societies that were functioning at the level of states, the most important of which at the time was the Aztec empire. In *Guns, Germs and Steel*, Jared Diamond gives a lucid account of why the Spanish military victory over the Aztecs was so decisive and swift. It was a pattern to be seen again and again in European–Indian encounters.

The Aztec empire was not even a century old when Hernando Cortez entered its principal city of Tenochtitlán in late 1519. Although sophisticated on many levels, Aztec society bore the unmistakable hallmarks of rapid growth through conquest, and the great majority of its subjects were conquered, controlled and resentful. One source of resentment was a constant demand for sacrificial victims. Their ceaseless flow kept Tenochtitlán's 5000 priests so busy that the conquistador Bernal Díaz del Castillo estimated the number of human skulls he saw piled with such regularity around the places of sacrifice 'at more than one hundred thousand. I repeat again that there were more than one hundred thousand of them.'[4] Another source of ill-feeling was the vast amount of tribute the conquered peoples were forced to yield. In one year, just before the Hispanic conquest, almost seven million kilograms of maize, 3.5 million kilograms of beans and amaranth, around a million kilograms of cotton cloaks and smaller amounts of arms, precious stones and feathers flowed into Tenochtitlán.

Many of these conquered people, together with urban commoners and rural peasants, were called *macehualli* by the Aztec. They were labourers who were bound to their place of residence much as were serfs in Europe, the Aztec lords inheriting them along with their land. Even among the ruling Aztecs, social organisation was hier-archical and involved tribute payments. Tenochtitlán was divided into four quarters, each divided into twenty *calpultin* (neighbourhoods), which were in turn divided into numerous streets. The system was inflex-ible, yet it allowed for efficient administration of a vast population. Across all of these divisions a rigid class system operated, with each *calpulli* paying tribute to the state.[5] These complex administrative struc-tures were maintained by the new conquerors to manage their own empire.

The Spanish also recruited members of the Aztec ruling class and relied on Aztec tribute books to determine what regions could yield which resources. In one famous example of this melding of cultures Cortez's Indian mistress Malinche bore him a son. Don Martin Cortez, one of the first Mexicans, went on to become a high-ranking govern-ment official. The extent to which the Spanish took over the Aztec empire simply by decapitating it and inserting themselves at its head can be seen in their rebuilding of Mexico City. A cathedral was built on the ruins of the main Aztec temple and the viceroy's palace occupied the site of Montezuma's residence, the symbolism of which would not have gone unnoticed by the Aztec survivors. From the perspective of the subject peoples, and if we set aside the ravages of disease, Cortez's conquest ultimately might have meant little—they were simply chang-ing one overlord for another.

To take greater advantage of their new conquests the Spanish revived a feudal relic called the Encomienda system. The conquistadors used it to grant themselves the right to collect tribute and to monopolise Indian labour, and it grafted almost seamlessly onto the old Aztec system of *macehualtin.*

Not surprisingly, given this history, the Spanish and Aztec spheres of influence overlapped and focused on the dense population centres that were vital to their continuance. The northern outposts of Spanish settlement, which lay beyond these centres, were usually subsidised or were marginal missions or government out-stations, whose cost was borne in part from Mexico. The most northerly clung to a fragile existence on the outer bounds of civilisation until later, when mixed Spanish–Indian families became ranchers and established a permanent and sometimes prosperous presence, in places as far afield as California and Texas. Their success, however, came about only through a long period of cultural adaptation that built on the strengths of two very different cultures, and a shift away from the colonial system built by the conquistadors.

In ecological terms we can think of the Spanish and the Indians as entering into a partnership. That partnership, which included aspects of parasitism as well as symbiosis, was one in which the Spanish ruled while most Indians toiled. Although the Spanish seemed to get the better part of the deal, it was to have one enormous drawback for them—it limited their control to areas peopled by settled agriculturalists living at high density, meaning the expansion of the Hispanic frontier was limited. This is not to say that the Spanish stayed timidly in their corner of the continent. Several expeditions, including Coronado's, which discovered the Grand Canyon in 1540, and that of de Soto, which traversed a vast band of territory between Florida, South Carolina and New Mexico from 1538 to 1543, contributed greatly to the exploration of regions that under a different system would yield enormous wealth.

In its early phases the French colonisation of North America was like the Spanish in that it developed as a partnership between old inhabitants and new. The French established settlements in the far north, near the northern limits of agriculture where the land abounded with fur-bearing mammals. It was in these mammals that the French would find their frontier. Their self-defined niche was one of middlemen in a trans-atlantic fur trade, at one end of which lay Huron hunters and trappers and at the other the furriers of Paris.

The French developed an exceptionally close partnership with their Indian trading partners. Theirs was a far more equitable, symbiotic relationship than that of the conquistadors and the Aztecs; the French frontiersman La Salle remarked, 'the savages take better care of us French than of their own children; from us only can they get guns and goods'. French traders often lived alongside Indians and intermarried with them; in effect they joined them in both love and war, for as well as fathering an army of mixed-race children, the French led many militant expeditions against the enemies of their Huron allies.[6]

Unfortunately, the French fur frontier was quickly exhausted. It burned like a wildfire across North America, forcing traders to move ever further west, south and north in their pursuit of furs. By the mid-eighteenth century the French had ventured deep into the continent, establishing trading posts and occupying much of the land earlier discovered by Spanish explorers, and making the wild west French long before it ever was the realm of the outlaw and cowboy. In the process they almost encircled the English settlements that were sprouting up along the eastern coast. Their occupation, however, was thin and tenuous.

The rapid French expansion and their close trading relationship with the Indians escalated conflict with the English settlers, who considered the selling of weapons to the Indians as the blackest of crimes. The deep ties between Indian and European Canadians continued long after the frontier period, for many Indian groups persecuted in the US during the nineteenth century fled to and found refuge in Canada. The declaration of a Canadian Inuit province in 1999 is perhaps the result of a greater mutual understanding and tolerance between Native Americans and European settlers than exists further south.

The French built their first permanent settlement in the New World on the St Croix River by 1604, and in 1608 they founded Quebec. The first successful English settlement in the New World was Jamestown,

Virginia, established in 1607. The Indians seemed to sense almost immediately that the English had a different sort of enterprise in mind, for when John Smith captured a Hassininga leader in 1608 he told Smith that the Indians had fought so hard because they believed that the Europeans were from the underworld, and that they had come 'to take their world from them'.[7]

By the time the pilgrims stepped ashore and began constructing their rude shelters at Patuxet in January 1621, Mexico City was an established and elegant European-style capital, its university seventy years old. Its cathedrals, markets and mansions were magnificent, and Mexican-Spanish influence had spread as far north as Florida and New Mexico.

At first the English looked to be inept colonisers, for their mortality rates were fearful and profits were meagre. What would eventually make them great was the character of their frontier—theirs was not to be a frontier of vassals or fur, but of the soil; and he who controls the soil controls the Earth's productivity, life itself.

The North American soil frontier was so bountiful principally because it was so extensive. Over the millennia the continent had laid up prodigious stores of soil and at no time was it generated in such large quantities as during the ice age. Louis Agassiz called glaciers 'God's great plough' because they gouged up countless miles of countryside and transported and crushed billions of tonnes of rocks, generating masses of fresh, new soil. Glacial streams and strong winds redistributed this bounty across the continent's more temperate reaches, laying down soils that nineteenth-century North Americans believed were created to await the arrival of their own more humble ploughs.[8]

So important was this glacial effect that the Australian geographer George Seddon once quipped that there is only one question you need ask of a continent in order to determine the fate of its people: 'Did you have a good ice age?' Seddon reasons that the ice age is *the* vital factor in determining the 'wealth of nations'—at least as far as the wealth of its soil goes. Because of its prodigious ice caps, North America had the very best of ice ages.

The implications of the English frontier for the Indians were to be dire, for by and large these settlers viewed Indians not as valued trading partners or a resource to be exploited, but as competitors. They were, to use a singularly American term, 'varmints', and they were chased from their land like any wolf or coyote. As a result, today the US retains less of its Indian culture than any other North American nation. All this, however, was far in the future on that winter day at Patuxet in 1621 when the first pilgrims began constructing their shelters; and before multiple and varied immigrations would depart England for the New World.

TWENTY-ONE

ENGLISH COLONIES ALL

One fine fall Saturday in 1998 I set out on a pilgrimage to Plymouth, New England, to pay homage to Patuxet Wampanoag and pilgrim alike. The place is quite modest, a pretty coastal drive past tacky gift shops sporting US flags, but boasts a grand granite edifice near the pier. The structure—a sort of cupola supported on sturdy columns—was erected by the 'Society of Colonial Dames' in 1921. It shelters a boulder dumped there by a glacier some 15,000 years earlier, and an inscription proclaiming it to be the spot where the nation's founders stepped ashore.

A little way off floats the *Mayflower* tied snugly at a pier, and above all is the hill where the pilgrims first started to build in 1621. The site is now occupied by an oversized bronze statue of Massasoit, the Wampanoag sachem (chief), erected by the 'Improved Order of Red Men', and an imposing granite sarcophagus reputed to hold the bones of the pilgrims who succumbed during that first terrible winter. A sign explains that the bones had been eroding out of the hillside for years and had been tossed into the cupola before being transferred to a more appropriate resting-place. The sarcophagus is inscribed with earnest prose relating how

> under the cover of darkness the fast dwindling company laid their dead levelling the earth above them lest the Indians should learn how many were the graves. Reader! History recalls no nobler venture for faith and freedom than that of this Pilgrim band [who] laid the foundations of a state wherein every man should have liberty to worship God in his own way.

It seems certain that at least some passengers on the *Mayflower* would have been horrified by this epitaph, imputing them to be supporters of religious tolerance. Like the *Mayflower* by the dock, Plymouth Rock and possibly the bones in the sarcophagus, the inscription is a figment of a later American imagination. It turns out that no one knows what the *Mayflower* looked like; the ship at Plymouth pier is an educated twentieth-century guess. Plymouth Rock itself is a dubious monument. In 1741 it was pointed out as the place where the pilgrims stepped ashore by ninety-five-year-old Thomas Faunce, who claimed to have heard the story from his father when a child. Even if Faunce had an exceptional memory, the sad truth is that his father arrived in Plymouth three years after the supposed event took place. The rock, historians suspect, is a crock. As for the bones in the sarcophagus, they may just as well be those of the Wampanoag who succumbed to a plague in 1617.[1]

As I pondered this fabricated history I wondered why the American people chose to elevate the rather straitlaced settlers to the status of fathers of the United States of America. A Harvard historian I asked ascribed it, somewhat facetiously perhaps, to the effects of the civil war. The victorious north, she believed, could not accept that the US was founded in Virginia in 1607. I felt that there was more to the matter than that, for the beliefs and ethics of both the pilgrims and the Puritans still have a place in the hearts of many of the country's citizens. There is clearly something in their values that has helped make the US great, but what?

Before coming to the New World many pilgrims lived in the Netherlands (first in Amsterdam, then in Stink Alley, Leiden) because it was then the only country in Europe tolerant of religious dissent. The Netherlands had a densely occupied human ecology where only the lowliest niches were open to unskilled, non-Dutch-speaking immigrants. Despite such difficulties, and its evident olfactory challenge, some pilgrims preferred Stink Alley to Patuxet, and their fate was to be the reverse of those who sailed, for they lost their culture and were transformed into Dutch men and women.

Many of those first settlers were a downtrodden and oppressed people with few of the skills necessary to make it in a pioneering society. Pilgrims from England and the Netherlands had been contracted to establish a settlement in the New World by the Virginia Company, whose sole purpose was to profit by trade. They should have been warned of the difficulties facing them by news of earlier English attempts at settlement in the New World. In 1587 all 117 members of the 'lost colony' of North Carolina vanished, leaving just the word 'Croatoan' carved on a tree. The first successful settlement at Jamestown did little better—in its first year in 1607 sixty-six of the original 104 colonists died.

In the early days of international capitalism human life was cheap, and the pilgrims were essentially company cannon fodder. Poorly supplied and equipped, they were dispatched at the worst time of year. When the advance party stepped ashore at Patuxet they were hungry and their clothes were frozen stiff. The main party disembarked on 2 January, at the beginning of a hard winter. Lacking shelter and sufficient food and clothing, they died like flies throughout January and February of 1621, and within six months half had succumbed. As is usual in highly stressed human groups, the children, women and single men suffered most, just a handful surviving.

The immigrants seem to have had no idea how to make a living in their new home. Were it not for a cache of Indian corn they found they would have all perished, for their supplies were inadequate and their English wheat and peas 'came not to good'. They were rescued from a dismal fate in March by Squanto, a Wampanoag Indian who had visited England in 1614, and stayed there for four years. On returning home in 1618 Squanto found that his entire tribe had died from disease the previous year. The epidemic had almost certainly been contracted from visiting Europeans.[2]

The Wampanoag, incidentally, formed the basis for the first report of an Indian culture from what is now the United States. Written by Giovanni da Verrazzano, a member of a French expedition that explored the New England coast in 1524, the report reveals a vibrant, numerous and fearless people whose villages and fields dotted the landscape. Da Verrazzano's account concludes with a long dirge sung by the Wampanoag for their dead, a fitting introduction to the scene of devastation that was to greet Squanto less than a century later.[3]

The pestilence of 1617 must have been particularly virulent, for it exterminated its entire host population. It had visited neighbouring Indian groups also, but just this one bit of coast was completely cleared of its inhabitants. This was a decisive factor in the success of the otherwise ill-fated Plymouth plantation, because it gave the pilgrims an opening in which to plant their settlement, yet left neighbouring Indian groups, with whom they could trade, relatively intact. This was just about their only piece of good fortune that first winter but it was a decisive one. Had their settlement been resisted by Indians anxious to protect their homeland, the pilgrims must surely have departed or perished.

According to the chronicler William Bradford, Squanto became the colony's lifeline in the spring of 1621. He directed the survivors how to plant corn, and showed them both the manner how to set it, and after how to dress and tend it. He also told them, except they got fish and set with it in these old grounds it would come to nothing. And he showed them that in the middle of April they should have store enough come up by the brook which they began to build, and taught them how to take it…where to take fish, and to procure other commodities, and was also their pilot to bring them to unknown places for their profit.[4]

Squanto's advice about burying a herring at the base of each corn plant illustrates just how rich New England's fisheries then were. Inexplicably, while the abundance of fish in New England waters was well known to both the pilgrims and their financial backers, no fishing equipment was carried on the *Mayflower*. Even more calamitous, the pilgrims had no idea about how to embark on the fur trade, which constituted the only other valuable commodity in the region. 'Neither was there any among them,' Bradford records, 'that ever saw a beaver skin till they came here and were informed by Squanto.'

Despite the difficulties they experienced, the shift across the Atlantic resulted in one enormous benefit: it removed the pilgrims from competition with other Europeans. In Europe they were at the bottom of the heap, persecuted and reviled as fanatics. Here they were masters of their own destiny as well as potential masters of a new land. The effects of this ecological release were felt from the earliest times, providing this transplanted society with an opportunity to fulfil its dreams. No one articulated these dreams as succinctly as John Winthrop, leader of the Puritans, who came to establish what is now Boston in 1630. The Puritan *raison d'être* was to be as 'a city upon a hill'—to serve as a beacon—a template for all righteous men who wished to serve the Lord. In their new city the Puritans would abjure the temptations of this world, awaiting fixedly the Second Coming of their demanding deity. It was only when riches were thrust upon them that many settlers were forced to alter their views.

Twelve years after they first stepped ashore, William Bradford bemoaned that

> the people of the plantation began to grow in their outward estates, by reason of the flowing of many people into the country. By which means corn and cattle rose to a great price, by which many were much enriched and commodities grew plentiful. And yet in other regards this benefit turned to their hurt, and this accession of strength to their weakness. For now…there was no longer any holding them together.[5]

The people of Plymouth were beginning to move away from the settlement. This group, 'so long together in Christian and comfortable fellowship', would never again meet as a whole in public worship. 'And this I fear,' Bradford lamented, 'will be the ruin of New England, at least of the Churches of God there, and will provoke the Lord's displeasure against them.'[6] In time the settlers would find their fortunes in cod and trade, but this first church-ruining fortune came, as so many later ones would, from a boom economy driven by a wave of Puritan immigration.

Between 1621 and 1640 more than 20,000 immigrants arrived in New England. With the rise of Cromwell, however, immigration halted and many young men returned to England to fight the cavaliers. Those who stayed behind embarked on a long period of comparative isolation, and began to define a strange new society. They formed an odd liquor, drawn off the European brew, then allowed to ferment in its own juices until the great migrations of the eighteenth and nineteenth centuries brought new blood and new ways of thinking to New England.

Within eight years of the cessation of immigration the religious fanaticism of Puritan society had become extreme. By 1648 you could be strung up for idolatry, blasphemy, man-stealing (a biblical crime), adultery, perjury, cursing a parent (if over age sixteen), third offences of burglary and highway robbery, or for being 'a rebellious son'—all of these in addition to manifold hanging offences brought over with the laws from England.[7]

The ferocity of their legal code was in part a response of the Puritans to their new home, for they believed America to be a stronghold of the devil. Everything about the place seemed God-forsaken, from the Indians (whom they took to be devil worshippers) to the untamed landscapes and wild creatures. A 1642 court case recorded by the long-suffering William Bradford reveals how deeply these beliefs worried them. He reported that 'wickedness did grow and break forth here', a wickedness so horrible that it precipitated a deep debate among the elders of the community. The wickedness did not involve a murder or robbery as one might expect, but sex. A young man, Thomas Granger, had been caught red-handed, so to speak, with a filly—of the equine type. Upon being questioned he admitted to having also been intimate with 'a cow, two goats, five sheep, two calves and a turkey'.[8]

The Puritan court focused on whether Granger had conceived of the idea for his barnyard bacchanalia in New England or old. This was of supreme importance to them, for many were terrified that 'Satan hath more power in these heathen lands'. Fearing for their own faith, they were greatly relieved when Thomas Granger admitted to having learned the vice from two inveterate horse-buggerers, both new arrivals from merrie England, where they had enjoyed their peculiar pastoral pleasure for years. So horrid was this crime in the eyes of the Puritans that they decided that the beasts must be punished as well as the lad, in the biblical manner. They had a problem though, for 'the sheep could not so well be known by his description of them'. Therefore 'others with them were brought before him and he declared which were they and which were not'. Bradford tells that 'a very sad spectacle it was. For first the mare and then the cow and the rest of the lesser cattle were killed before his face…and then he himself was executed' and the corpses cast into a pit.[9] Given their fears of the devil's power in America, the Puritans

inexplicably failed to punish the only true American in the whole affair—
the turkey. Perhaps some devout soul had already enjoyed it at
Thanksgiving!

The Salem witch trials of the early 1690s did much to destroy
religious tyranny in New England, but for decades after Puritan ethics
ran deep. Puritan culture was the outcome of enormous societal stress,
for it developed during one of the most uncertain and calamitous periods
in English history. Its emphasis of self-denial and extreme conformity
with an uncompromising faith seems typical of social movements
that arise at such times; whether contemporary US Fundamentalist
cults or religious fanaticism in the Middle East. Societies founded on
such principles are like glass—at once hard, yet easily fractured. They
thrive during times of adversity, but during more prosperous periods
their stringent belief systems may be either discarded or greatly modified.

The founding fathers present a profound engima to the ecological
historian who imagines that he might find in them an equivalent of the
first Clovis pioneers, or at least a people keen to embrace the bounty of
the frontier. Instead one finds the opposite. Although they grew corn in
Indian style rather than wheat, and husbanded (evidently on occasion
a little too literally) turkeys as well as chickens, their ecology was not so
different from that of villagers in the old country. Thus, at an ecological level, despite the shift of continents, the Puritans continued to live
very British lives. Even today it is a commonly held view that New Englanders are the most European-like of all US citizens.

For the Puritans the frontier was not so much a magnet as a barrier,
or at least a bedevilled land. The western limits of their settlement lay
along the Connecticut River valley, beyond which was different country,
unfriendly to the English agriculturalist. For a century and a half the
Puritans ignored the call of the land beyond this frontier, unwilling to
change in a manner that would permit them to make a living in the
country lying to the west.

This may be one reason the frontier failed to develop, but I suspect another lies in the fact that New England was not sufficiently remote from the old. A sea journey of a few weeks was all that separated Boston from London, and that is a very different thing from a trek of several months through the wilderness. The real America—the land of the frontier—would arise from other beginnings.

In 1724 the Reverend Hugh Jones astutely summarised the various English colonies that had sprung up on the Atlantic coast as follows:

> If New England be called a Receptacle of Dissenters...Pennsylvania the Nursery of Quakers, Maryland the Retirement of Roman Catholics, North Carolina the Refuge of Run-aways, and South Carolina the Delight of Buccaneers and Pyrates, Virginia may be justly esteemed the happy Retreat of true Britons and true Churchmen for the most part; neither soaring too high or drooping too low...[10]

Jones's characterisation of the Virginians as 'true Britons' is curious, because the English who settled there had devised an extraordinary new and very un-English means of making a living. Indeed they had shifted into a new ecological niche, one that must have appeared completely alien to the English agriculturalist of the day. That niche was made possible not by standing aloof from their new home, nor by entering a partnership with a New World people, but through an ecological alliance with a peculiar New World weed and another Old World society: for the Virginians, slaves and tobacco would shape the future.

In order to prevent their leaves being eaten by insects, the shrubs and herbs that make up the genus *Nicotiana*, to which the tobacco plant belongs, produce an extraordinary cocktail of chemicals. On average 10 per cent of the plants' metabolic effort is spent producing just the alkaloids that go into the mix. The flow of chemicals from the glands of some American species is so copious that it literally drowns the insects attempting to make a meal of it. Such devotion of its metabolic efforts indicates that the plant faces formidable enemies in nature, and tobacco-eating insects are a resilient lot. Indeed, the newly hatched larvae of the tobacco grub recoil at their first bite of a tobacco leaf, but soon reconcile themselves to their toxic food and will thereafter take no other.

South America is the principal home of the tobacco genus, but some members are flung as far and wide as southern Africa and Australia, suggesting that the lineage may be a venerable one which evolved at a time when these landmasses were joined. The North American species probably travelled north with the glyptodonts and sloths after the formation of the Panamanian land bridge. People, like tobacco grubs, acquired a taste for the toxic chemicals, and it was from among these immigrant species that North American Indians first selected plants that offered a good smoke. By AD 500 the Maya were already in the habit, and by the seventeenth century smoking had spread through vast areas of the continent.

Curiously, the tobacco enjoyed by Sir Walter Raleigh and first grown in Virginia in 1612 is not the species cultivated for smoking today. Raleigh relished the aroma of *Nicotiana paniculata*, which, although it is no longer smoked, has not entirely vanished as a crop—it is still grown in Eurasia as a source of insecticides. The plant that fills the fields of the American south today is *Nicotiana tabacum*, which appears to be a hybrid species with Argentinian and Bolivian ancestry.

Tobacco chews through soil fertility with a ferocity possessed of few other crops, for it needs rich soil to keep up its chemical defences. In the absence of fertiliser it can be grown successfully only on virgin land. The second crop is the best, but after four seasons the land must be abandoned to less demanding crops such as corn. When its usefulness for even these crops was finished, the Virginian colonists allowed their land to revert to secondary forest. This peculiar ecology indicates that tobacco is a pioneer species—one of those plants that waits to spring up in freshly bared soil. Its unique characteristics meant that tobacco could only be grown where both virgin forest and the vast amount of labour needed to clear it were to be had cheaply. The labour was to come in chains from Africa, the land to be wrested from the hands of Indians.

Tobacco was a pioneer plant destined to give rise to a pioneer culture, and for the sake of tobacco Virginian agriculture and settlement became semi-nomadic, following the newly cut frontier ever westward. This caused the centre of population to move with it, forcing the founding and abandonment of a succession of capitals—first Jamestown, then Williamsburg and finally Richmond—each one further west than the last.

Save for its seasonally inhabited capitals, Virginia was without significant townships until well into the eighteenth century because most people lived on plantations. The southern plantation finds its closest analogy in the modern world in the company town, for a large plantation had 'hundreds of slaves, white craftsmen, overseers, stewards, and traders who were producing tobacco as a money crop, raising food, and manufacturing tools, farm instruments and clothing' all of which could be used locally or traded.[11] Here was a social and economic unit the likes of which did not exist in seventeenth-century Britain. It was an original adaptation to life in the New World that had been driven by the needs of a New World crop. So demanding was the weed that the plantation system may well represent the first great experiment in large-scale commercial agriculture since the days of the Roman empire.

While the Virginians had, at least in an ecological sense, travelled far towards becoming a distinctively American people, they had remained, as the Reverend Jones observed, deeply British at the social level. Their adherence to the English church and English customs partly derived from the fact that they had come to the New World to seek their fortune, many intending (at least initially) to return home when they had made good. Thus they clung to the majority faith and remained English in their retention of a hierarchical society. The egalitarianism that is such a feature of nascent European American culture elsewhere was conspicuously lacking in colonial Virginia, a situation doubtless encouraged by slavery.

Through the eighteenth and the first half of the nineteenth century slavery became ever more rigid and institutionalised in America. The plantation system also diversified and tobacco's supremacy was challenged by other crops, the most important of which, a member of the hibiscus family, soon gained the name 'king cotton'. Members of the cotton genus (*Gossypium*) have an almost pan-tropical distribution, and were domesticated independently in both the New and Old Worlds. The uses to which the plants have been put seem almost endless: from production of oil and flour in America, to yellow dye in India and manufacturing of a male contraceptive in China. It was the fact that the plant literally keeps its young wrapped up in cotton wool, however, that made it significant to the plantation system. Its seeds are covered in single-celled hairs, each 3000 times as long as wide, which flatten when they dry and can be spun into cloth. The southern planters used American varieties to stock their cotton fields, and this American plant would prolong slavery in the New World.

As the eighteenth century wore on slavery became increasingly odious in the eyes of Europeans, and the Virginian aristocracy became caught in a cleft stick. They identified themselves as British gentry and made enormous efforts to 'keep up appearances'. With their enthusiasm for such noble traditions as fox hunting, they even introduced red foxes from England to North America, despite the fact that the continent already had perfectly good red foxes. They returned 'home' to England often, but there were increasingly treated as country cousins and immoral owners of slaves. Over time, the fundamental incompatibility of the Virginians' ecological niche with its reliance on slavery and their social identification with the English gentry would become more and more of a problem.

To summarise, we can imagine the response of the Puritans and Virginians to their new homes as diametrically opposite. For the Puritans their way of life was English, but their emotional home was America. The Virginians, on the other hand, were living and farming as no one had before. In an ecological sense they had become truly North American and were poised to open up the frontier. Yet despite this ecological revolution many Virginians were, at an emotional level, profoundly English. They saw themselves as the king's most loyal subjects and, although they wanted change, revolution was something they preferred to avoid. Just how the Puritans and Virginians combined to form the United States of America reveals unexpected strengths and weaknesses in their different ecological and social responses to the New World.

The new nation was born out of two remarkable movements—a revolution and an act of union. Over time most colonies have sought independence, so revolution comes as little surprise. Successful union, however, is a much rarer event. The historian Frederick Jackson Turner believed that the American union was remarkable enough to deserve further study, writing a century ago that in it 'lie topics for the evolutionist'.[12]

Similar efforts at union elsewhere in the Americas have failed or proven to be ephemeral. Simón Bolívar's dream of a United States of South America remains unfulfilled, while the more recent proposal for a federation of Caribbean states died at conception. Even Canada, which achieved nationhood and federation in 1867, is now experiencing a reverse, for Quebec seems destined for independence, while in 1999 the Inuit achieved statehood and autonomy. There are powers in the lives of people that both encourage and dissipate confederations, and the United States is fortunate in being blessed predominantly with the former. Successful federation may well be so infrequent because if it is to work it requires that politicians relinquish some power—a rare occurrence indeed in the real world.

Intriguingly, the impetus for the American union may have come in part from indigenous America, for it has recently been argued that Indian political organisations, particularly the sophisticated Iroquois League, influenced America's founding fathers. Benjamin Franklin heard Canastego, the Onondaga sachem, speak to the commissioners of Pennsylvania, Virginia and Maryland on the subject of unification and its benefits. Yet Indian influence can only be part of the story. What else could have led the very different English Americans to join together, and what would the new, syncretic culture be like—dour New England or the fun-loving, genteel south?[13]

The American War of Independence was a bitter struggle, a true civil war whose capacity to destroy the fabric of society is easily underestimated today. Almost as many Americans (some 8000) fought with the British to preserve the *status quo* as fought in Washington's army (about 9000) and the issue divided many families; in the case of the Franklins it pitted father against son—a by no means atypical even if tragic outcome. By the war's end 80,000 to 100,000 people had emigrated, mostly to Britain, Canada or the West Indies.[14]

The war soon made other divisions apparent, for no one was allowed to remain neutral. Almost all American Indians who fought took the side of the British, as did 800 slaves. Both had something to gain from British rule—the Indians supporting a law limiting European colonisation to the east of the Appalachians, while the slaves were offered freedom in return for military service. So bitter did the conflict become that barbarities were committed on both sides, including the execution of prisoners without trial. It was in the context of this enormous societal stress, with families turned against themselves and with colonists leaving *en masse* for Canada or Britain, that the first pan-American culture was born.[15]

The decision to do away with king and old country and to embrace republicanism was a momentous one, and unity was a prerequisite to revolution, for those colonists that led the revolt opened themselves to potentially fatal consequences. Treason was a hanging (and drawing and quartering) offence and Benjamin Franklin warned, 'We must all hang together, or assuredly we shall all hang separately.' Even given the imperative to 'hang together', achieving unity was no easy matter. By 1774 great differences had emerged between the colonies and no greater differences existed than those between the New Englanders and the Virginians. As the ever observant Reverend Hugh Jones noted, 'the common planters [of Virginia]…don't much admire labour, or any manly exercise, except horse-racing, nor diversion, except cock-fighting, in which some greatly delight'. These activities, along with other beloved southern pastimes, gambling and the theatre, were forbidden in the puritanical colonies to the north. Not surprisingly the Virginians looked upon their northern neighbours as dangerous, sour-faced wowsers in matters of religion and politics. Even fifty years after the revolution, Fanny Trollope could write, 'if you hear of his [New Englander] character from a Virginian, you will believe him a devil'.[16]

When the first Continental Congress convened in September 1774 it faced the task of building a unified nation from these disparate parts. It began its work, paradoxically, by banning many of those pursuits considered corrupt and debilitating in New England but which were loved in Virginia, such as cock-fighting, horse-racing and the theatre. In a stunning understatement one historian commented that 'the New England delegates to the congress had to guard against giving the impression that New Englanders stood aloof on a pinnacle of purity'.[17]

It appears that New England Puritanism served as a battle cry for national unity because it offered a suitable avenue for protest and a show of solidarity during these infant hours of uncertainty. Puritanism was an outgrowth of fractured times, when denial and self-abnegation are typical human efforts to mollify the gods. The colonists' sacrifice not only forged a compact with God, it made a political statement. It set them apart from the British as a self-proclaimed pure and determined people (even if they were a little sour-faced). Most importantly it announced that their claims were just and had to be taken seriously.

The triumph of puritanical nationalism was not without effort, for that singular American innovation, tarring and feathering, was practised with enthusiasm in order to bring about conformity with the new laws. Mercifully, capital punishment was not resorted to until the outbreak of war, and even then only royalists actively helping the British were turned off the scaffold.[18]

By April 1775 war with Britain finally broke out. The year was memorable not only as the beginning of the American War of Independence, but because it marked the arrival of the first brown rats (*Rattus norvegicus*) in North America. These pests, it seems, had reached the continent aboard English ships. England herself had received them from Norwegian timber vessels sailing from Russian ports as late as 1728–29. As a result of a coincidence in time between its appearance in England and the coronation of the first Hanoverian monarch, the pest was dubbed the Hanoverian rat. This rodent was to become a scourge in the stores and docklands of North America, extracting its tithe of all perishable goods that passed its way. It ensured that the Hanoverian 'Mad' King George III would have his 'taxation without representation' after all.

The close of the eighteenth century saw the United States a unified nation, yet its inhabitants still clung to the region east of the Appalachians. To the south, Mexico was still a dominion of Spain and would remain so until 1821. To the north Canada was a newborn and, in terms of its population, an insignificant British colony. Its ranks, however, had swelled with a mass immigration of English loyalists from the newly independent United States. Independence for Canada would come peacefully in 1867, the same year that Prince Gortchakoff announced that Russia would sell Alaska to the Americans for US$7,200,000 in gold.[19] The New World states were still inchoate, but even then were beginning to show signs of greatness.

On the ecological front, by the end of the eighteenth century the European Americans had exterminated the most desirable game such as beaver, bear, deer and turkey from southern New England. They had not, however, eliminated any of North America's native species. They were not yet a continent-transforming people, but in that revolutionary year of 1776 Daniel Boone laid out the Wilderness Road across the Cumberland Gap and Kentucky, and the rich western horizons lay open.[20]

Early on a beautiful spring morning I made a pilgrimage to Monticello, the home of my favourite North American, Thomas Jefferson. I decided to walk up the high hill to the house rather than drive. The path took me past groves of pink bud holding forth their blatantly sexual promise of new life, past great oaks that must have been substantial trees in Jefferson's time. I climbed past the family cemetery where the great man lies buried, past the vegetable gardens where his slaves (many of whom seem to have been his extended family) worked, and on to the hill where Monticello stands, with its views of Charlottesville and the Blue Ridge Mountains beyond.

Jefferson designed the building and its construction was the work of a lifetime. It has an amazing modest and compact beauty, revealing the Jeffersonian mind as nothing else does. It was at the mansion's door that I saw the first of his famous devices—an ingenious extension of a weathervane allowing one to read wind direction without going outside. Just inside the door was another engineering marvel, a seven-day clock, while beyond lay sliding doors, windows, beds and staircases all designed by the great man and indicating that the American romance with machines has long roots indeed. This love of machinery also had its practical side, for Jefferson, like most plantation owners, lived a life isolated from town and city and, for example, could not easily call in a watchmaker if his clock broke down. Better to design, understand and repair the mechanism oneself. Here I found my first echo of the self-reliant frontiersman. Indeed, in Jefferson's time the frontier lay just across the Blue Ridge Mountains, so beautifully beckoning from his high hill.

To enter Monticello is to be transported back to the time when the frontier had just begun to open. In the foyer the fossilised bones of a mastodon that had been excavated nearby are displayed. Jefferson believed that somewhere beyond those beckoning Blue Ridge Mountains mastodons still existed. Standing in that foyer I could still sense the excitement, the anticipation roused when he finally obtained the funding to send Lewis and Clark over those mountains and on to the distant sea. The mementoes that they returned with—the Indian clothing, the buffalo skin and antlers of elk—still adorn the house, and they still speak to me of that magical time when the west was not yet won.

TWENTY-TWO

CONCEIVED IN LIBERTY

Make way...for the young American Buffalo—he has not yet got land enough; he wants more land as his cool shelter in summer—he wants more land for his beautiful pasture grounds...We will give him Oregon for his summer shade, and the region of Texas as his winter pasture. Like all of his race, he wants salt, too. Well, he shall have the use of two oceans—the mighty Pacific and the turbulent Atlantic shall be his.[1]

So did a delegate at a Democratic state convention in 1844 express the national mood as the wave of frontier expansion began in earnest. It was a mood of unbounded enthusiasm and confidence in the face of which neither soil, nor Indian, nor buffalo would be safe. Yet for all their enthusiasm for expansion, the North Americans would not fully understand how the frontier experience had changed them until fifty years later, three years after the frontier was officially proclaimed closed.

Chicago's World Columbian Exposition of 1893 is all but forgotten today. Tourists visiting the site see a vacant lot and a park, outside of which at a busy intersection stands an isolated, unexplained gold statue of a woman. 'All that's left,' tour guides say. 'She's supposed to represent progress or something.' At the time, however, the Exposition was a dizzying event, its eighteen hectares of arts and manufactories competing with innumerable food stalls, novelty rides and perform-ances, including Buffalo Bill's Wild West Show, which was a sell-out.

Amid the ephemeral glitter one seemingly trivial event took place that would have a profound impact on the way future North Americans would see themselves. It was a history lecture titled 'The Frontier in American History', delivered by thirty-two-year-old Frederick Jackson Turner from the University of Wisconsin. Turner began by quoting the superintendent of the 1890 census. 'Up to and including 1880 the country had a frontier of settlement,' the superintendent wrote, 'but at present the unsettled area has been so broken into by isolated bodies of settle-ment that there can hardly be said to be a frontier line. In the discussion of its extent, its westward movement, etc., it can not, therefore, any longer have a place in the census report.' The frontier, Turner announced, was over. Even though first-time homesteading claims would still be made in northern Montana until the 1920s, the young historian, speaking just three years after the massacre at Wounded Knee, had made a telling point.[3]

Turner may be out of fashion among historians today, yet he displayed great insight into the way that the changes originating on the frontier shaped American society. I believe that Turner was correct in seeing the frontier experience as a significant force in shaping the United States we know today. He viewed the first part of the process by which the frontier changed people as a sort of 'cultural stripping'. 'The wilderness masters the colonist,' he wrote. 'Complex society is precipitated by the wilderness into a kind of primitive organisation based on the family.' It was the harshness of frontier life that brought this about, for it was a world with the most basic amenities, with only rudimentary material goods and little news of the outside world. Education was haphazard, so much cultural information was not passed on down the generations. 'In the crucible of the frontier,' wrote Turner, 'the immigrants were Americanised, liberated, and fused into a mixed race, English in neither nationality nor characteristics.'[4]

This cultural stripping was an integral component of the frontier experience, for it took English, Yankees, Germans, French and a hundred other types, and converted them all into frontiersmen and women. As Turner recognised, the frontier had a huge homogenising influence, drawing in a plethora of cultural diversity and churning out a uniform type. These people, he claimed, were the first true Americans:

> The advance of the frontier has meant a steady movement away from the influence of Europe, a steady growth of independence on American lines. And to study this advance, the men who grew up under these conditions, and the political, economic, and social results of it, is to study the really American parts of our history.[5]

Turner argued persuasively that the frontiersman came to adopt Indian habits and ways of life. The frontier, he said, takes the European 'from the railroad car and puts him in the birch canoe. It strips off the garments of civilisation and arrays him in the hunting shirt and moccasin'. And the settlers were to become like Indians in other ways; nearly every colony established itself on the site of an old Indian village. They grew Indian crops such as corn and squash, and according to European observers they even started to farm like Indians. They put the turkey in their barnyards, and even became vulnerable to European epidemics. Most striking in this regard was their increased sensitivity to smallpox. The Reverend Hugh Jones noted that more Virginians would have received a continental education 'were they not afraid of the small-pox, which most commonly proves fatal to them'. Smallpox did not always wait for the colonists to travel to Europe. It swept through their settlements, sometimes at intervals of more than a generation. No longer just a childhood scourge, it became a terrifying force that paralysed whole communities, for smallpox epidemics could result in the suspension of all intercourse with the afflicted region, along with the deaths of a significant number of its citizens.[6]

Turner also argued that there was not one but several frontiers that followed one after the other, with each successive frontier exploiting more intensively the resources on offer. The first frontier was that of the hunter and trader, then came the pastoralist, the farmer and finally the city-builder. As each frontier moved west the process of transforming its people was repeated many times. The advance of the initial wave, the trading frontier, was so rapid that the pioneer often beat the explorer to the first glimpse of virgin land. Although Turner does not mention it, this frontier was largely French. In the next wave—the pastoral frontier—land fortunes were made and lost while, in the wave to follow, people gambled the soil on a mad intensification of cropping as if there was no tomorrow.

Turner's insights are so fundamental that they transcend the American frontier experience and help us understand frontiers anywhere. Indeed, the generalities of his hypothesis are as much organic and biological as they are social. His idea that the frontier consisted of successive waves finds an echo in Indian archaeology, beginning with Clovis the mammoth hunter, then Folsom the bison people, followed by rapid diversity: farmers, hunter-gatherers, clans, chiefdoms and city-building states. Here, however, the changes were not due to waves of migration but to *in situ* evolution.

Not surprisingly, given the frontier's ability to strip Englishmen of their culture, the eighteenth-century British government had a decided aversion to it, and attempted to limit its effect by containing all European settlement to the areas east of the Appalachians. Their stated reason for doing so was concern for the welfare of the Indians, whom they were obliged to protect, but the loss of control that the frontier implied also loomed large in their thinking. To limit expansion, however, was intolerable to many colonists, especially to those living in the south, for they were becoming a true frontier people. The determination of Americans to expand made British attempts at limiting the frontier useless. It was as if a biological phenomenon was in train, which mere laws and prohibitions could not control.

The striking thing about the soil frontier is the length of time it stayed open, for it was an important element in North American life for nearly 300 years. As a contrast, in Australia, which is about the same size as the contiguous forty-eight states, the land frontier lasted less than a century while, in any one location, nature's first bounty was often exhausted in less than a decade. This, I suggest, accounts for many of the great differences between American and Australian cultures. The North American experience of 'ecological release' encouraged its inhabitants to develop new ways of harvesting an almost unimaginable natural plenty, while the Australians found themselves facing adversity almost from the moment they entered the continent. Australian cultural evolution tells of a long struggle to exist within the limits that nature placed upon its people. In other words, they adapted to their new home, while North Americans were released by theirs.

'The growth of nationalism and the evolution of American political institutions were dependent on the advance of the frontier,' wrote Turner. Certainly warfare, driven by notions of manifest destiny, was a major feature of the frontier and was the principal mechanism by which the American colonies, then later the United States, expanded. The most important of all frontier wars were the Indian wars, which continued almost unabated between 1608 and 1890, and which delivered the lion's share of North America into British American hands. Wars with other European colonists were also frequent and helped expand borders. The

Seven Years War (the French and Indian War as it is known in the US) delivered the Ohio territory. The war of 1812 (the invasion of Canada) was less successful, but the war for Texas independence (1836), and the Mexican–American war (1846–48) delivered big territorial gains. This last war has been characterised by historian Richard Shenkman as 'a war for pure territory…not the kind of war Americans care to celebrate or mythologise, nor should they'. Still, it gained the US 1.3 million square kilometres. The Spanish–American war of 1898 was little better, but it gave the young nation its first overseas empire, which included the Philippines, Guam and Puerto Rico.[7]

Despite their hunger for land, Americans have not always kept what they conquered. In 1813 a US army burned York, the capital of upper Canada. They were forced to retreat, and in revenge the Canadian–British forces burned the White House. And in 1847 a victorious army raised the Stars and Stripes over Mexico City, yet Congress baulked at incorporating Mexico into the United States.

A century on from Turner, we all imagine that we know the kind of individual the frontier created: tough, self-reliant, uncultured, defiant of government and law, and standing alone against nature's adversity. Perhaps it did this, some of the time, but a different and far more significant change was engendered at the frontier. The very essence of the frontier experience lies in the extent of its resources, and when resources are boundless, why conserve them or even utilise them efficiently? The principal goal is to exploit them as quickly as possible, then move on. It is this frontier attitude to resource utilisation that lies at the heart of much capitalism, and which presents such a major challenge to conservationists today. In this sense, the legacy of the American frontier is still very much with us.

Humans, of course, were not the only living things to find a frontier in North America, for the Europeans brought a whole new megafauna with them, as well as countless other living things, and each would experience its own ecological release on its own frontier. Ecological release is a profoundly important factor in determining the fate of translocated animal populations. It occurs when a species' competitors, predators and diseases are left behind as it moves into a new and favourable environment. It is pervasive in long-isolated places such as Australia where foxes, cats, cane toads, rabbits, sheep and European humans—to mention just a few—have experienced it over the last two centuries, much to the detriment of the native fauna. The release is heightened if there is a 'naive' resource base that the new species can use to grow rapidly in numbers. While it is an important evolutionary process, ecological release is usually short-lived, lasting only as long as the species' 'frontier' remains open. Its transient nature means that it is difficult to study in the fossil record, but historical events can teach us much about the process.

The faunal interchange between Eurasia and North America that followed 1492 was different from all earlier exchanges in that it was mediated by humans. Many species, such as rats, mice and the small-pox virus were introduced inadvertently, but people deliberately selected and transported larger mammals and birds. As with all programs of introduction, far more species were transported and released than survived and spread.

Among the first and most important of immigrants to arrive with the Europeans was the horse. Its extinction 13,000 years ago left the continent horseless for the first time in over 45 million years. The first to breathe the air of the New World after this 13,000-year absence was carried across the Atlantic on Columbus's second voyage, arriving on the Caribbean island of Española in 1493. Ironically, the Caribbean islands were the one part of the Americas that had never been inhabited by horses.

Horses finally reached the continent of their evolutionary birth with Cortez, who took fifteen of them with him on his conquest of Mexico. They were Barbs; small, tough horses that originated around north Africa's Barbary Coast. Cortez noted the colour of some of these equine pioneers; black, grey, piebald. One colt, the first to be born in North America for thirteen millennia, escaped from the great conquistador and became the first wild horse to roam the continent since the Pleistocene. But it was a false dawn, and the solitary creature probably died a virgin, unable to fulfil its potential to repopulate the continent. Some twenty years later the de Soto expedition released six horses in what is now the southern US. All were near death when set free, so it seems unlikely that their progeny filled the plains either.

These lonely pioneers were soon followed by others, either stolen or lured from the Spaniards and which then escaped. Their numbers were greatly augmented by the horses that ran wild during the Pueblo Revolt, led by Popé in New Mexico in 1680. Meanwhile, larger horses of northern European origin were arriving on the Atlantic seaboard, reaching Virginia by 1620 and Canada by 1665. But it was the mustangs—descended from the Spanish Barbs—that would fill the continent. By the time of the War of Independence herds had become established as far afield as northern California. They arrived in the American northwest just in time to help the beleaguered Lewis and Clark who, hungry and exhausted, needed to cross the Rocky Mountains before retracing their steps back east.

By the early nineteenth century mustangs had become even more widespread than buffalo. From the deserts of the south-west to the swamps of the south-east and north to the Canadian prairie the land once again echoed to the whinnies of the wild herds. It had taken them just 300 years, around the same span it took the Clovis and British pioneers, to occupy the continent after a 13,000-year absence.

At some stage wild horses were joined by runaway burros (asses to Europeans), and by the end of the seventeenth century great herds of both were running free in the southern parts of North America. So plentiful were horses by 1820 that when a US Army remount officer told a hacendado that he required a thousand, the laconic Mexican's only response was 'Do you want greys or chestnuts?' By 1900 there were around two million mustangs roaming North America but, like the buffalo, their reign would soon pass. Hunted, harassed and ever more closely fenced in, their numbers were halved in the next twenty-five years. A decade after that just 150,000 remained, and by 1967 their numbers had dwindled to just 17,000. In 1971 legislation was passed to protect the last wild mustang and burro herds, but their wild west frontier had vanished, and they will once again have to learn to live with man.

Other Eurasian immigrants—this time botanical ones—would put the finishing touches on many American horse scenes. The 'Kentucky' bluegrass that has fed so many thoroughbreds would soon arrive from Eurasia, and by 1886 the tumbleweed would blow in from Russia. From then on, the mustang and the cowboy and his mount would wander the west in its company, completing that definitive image of the American frontier.

Today the ranges of the wild horse and burro in North America centre on the arid south-west where both species had close relatives 13,000 years ago. They should be regarded as reintroduced species following Pleistocene extinction rather than new introductions, and treated accordingly. Today feral burros are accused of competing with mule deer, American antelope and desert bighorn sheep (*Ovis canadensis*). 'The native desert bighorn sheep of both Death Valley and the Grand Canyon are doomed,' one report suggests, 'unless something is done to eliminate wild burros.'[8] The very language—'native' for the sheep, 'wild' for the burro—reveals something of the prejudices operating here. What we may be seeing is an instance of range stabilisation somewhat as it was in the Pleistocene, rather than extinction of a native by an introduced species.

The incredible history of the horse in North America raises a question—why did it become extinct so rapidly 13,000 years ago, and then thrive so marvellously following its arrival in the sixteenth century? Several factors must have played a role, but two may have most of the power of explanation. The first concerns the naivety of the Pleistocene American horses. None of these creatures would have recognised danger in a Clovis hunter, for humans look and behave like no predator that the pre-Columbian American horses would have been familiar with. It is more likely that they would, at least initially, have curiously approached the first hunters they met. The second factor is that 13,000 years ago horses meant only one thing to people—meat. When they returned in the sixteenth century they arrived as domesticates, and in the Spanish the Indians found a template of how to work with a horse. The lesson was not lost on them, and soon a living horse was worth far more than a dead one. A compact had thus been agreed between the species that would see the horse return to the continent with the help of humans.

The pig arrived in the New World at the same time as the horse. It proliferated on the Caribbean islands and by 1560 had become well established in Florida, then spread throughout much of the southern, eastern and western coastal regions of the United States. The family to which the pig belongs (the Suidae) has never naturally been present in North America. Prior to 13,000 years ago, however, the most common large mammal on the continent was a broad-headed, pig-sized peccary known as *Mylohyus*. It seems possible that the razorback may be partially occupying the ecological niche once occupied by this long vanished peccary.

Cattle and goats have also established feral populations, with cattle forming a partial replacement for bison. Following the demise of the bison in the early 1880s, free-ranging cattle flourished on the Great Plains. It was a short-lived glory, however, for fierce blizzards in the winter of 1886–87 destroyed the herds. Cattle living in paddocks have a very different ecological impact from those that roam at will, for they cannot migrate and so the impact of their grazing can be far more destructive than that of wild buffalo. Among the remaining less successful large, non-domestic ungulate introductions to the US are the axis, fallow, sika, red deer and the nilgai (an Indian antelope), many of which became established in Texas as a result of releases by the King Ranch Inc. All, however, remain localised in distribution.[9]

No European carnivore has succeeded as an immigrant into the New World except the red fox, which has occupied much of the eastern seaboard, an area where the native North American red fox evidently was absent. The European red fox has had no observable impact upon the native North American fauna, a situation very different from that in Australia where it is implicated in the extinction of about twenty mammal species.[10]

North American carnivorous species, in contrast, have been highly successful migrants to Europe. Foremost is the mink, which now occupies vast regions of Eurasia, from Ireland as far east as the Kamchatka Peninsula. The raccoon has also been successfully introduced to Europe, but its spread has been limited to parts of Germany and adjacent countries, and a few localities in what was the USSR.

Some smaller North American herbivores have also invaded Eurasia, including the muskrat, which has been hailed as the most successful feral mammal in the world. Beginning with the introduction of three female and two male muskrats to Bohemia by Prince Colloredo-Mannsfeld following a hunting trip to Alaska in 1905, this species has spread throughout northern Eurasia. It is now present from eastern France to Scandinavia, and eastward to Japan and the Kamchatka Peninsula. The area of its distribution in Eurasia now exceeds that of its homeland in North America. In addition, the Canadian beaver is found in far eastern and western Eurasia, the grey squirrel is present in Britain, while a few thousand white-tailed deer established themselves in the former Czechoslovakia and Finland.[11]

The overall nature of these introductions reveals more of the economic needs of the Europeans than it does of the ecology of the individual species involved. In North America the colonising Europeans needed large ungulates for transport, labour and food. Europeans at home, on the other hand, needed fur and thereby introduced and tried to farm mink, Canadian beaver and muskrat. Despite this economic bias the great exchange that began in 1492 offers us a laboratory in which to examine earlier mammal immigrations. The most striking feature of the historic introductions is that none has caused total extinctions among the natives. The American and European beavers differ slightly in their ecology and do not displace one another because they occupy different habitats. The burro has precipitated some harm on the bighorn sheep but seems unlikely to drive it to extinction throughout its range. In Europe the grey squirrel has had a large impact upon the red squirrel, yet the latter still survives in parts of Britain.

Over the past 60 million years, North America has rarely been free from new invaders from Asia or South America yet, as is the case with the historic introductions documented here, few if any extinctions are emphatically attributable to them. Where extinctions do occur in the fossil record they correlate with climate change more than immigration, and it is hard to escape the conclusion that climate change was the cause.

This pattern provides us with very valuable information, for it aids us in learning how to manage introduced mammal populations and anticipate the impacts of others. Most contemporary ecologists look on animal and plant introductions with disfavour, citing the very worst cases of damage to argue against such introductions. Yet these cases have usually occurred on long-isolated islands such as in the Caribbean and Australia. Extensive ecological damage as the consequence of an historic introduction of a large mammal has never occurred in North America. Given the fact that the continent has never supported a more impoverished mammal fauna in the last 50 million years than it does at present and that the existing fauna is unbalanced, appropriate introductions are more likely to be beneficial rather than deleterious.

The one great exception to this concerns our own species. It alone has caused massive extinctions on immigrating to the New World, not once but twice. As we have seen, the first humans to enter the New World appear to have exterminated most of North America's large mammals. The arrival of a new kind of human, with different technologies and ways of doing things, would repeat the catastrophe all over again.

Although the flow of new species of large mammals into the North American environment has now slowed if not ceased, other sorts of immigrants from Eurasia have begun to cascade in as new technologies have opened, for the first time in 50 million years, access for temperate and tropical species. Among the most devastating of all of these new invaders are the diseases of plants. A fungus that originally lived in Asian chestnuts was among the earliest to arrive. It was imported in 1904 in a shipment of nursery trees and soon swept through the chestnut population, killing hundreds of millions of trees and effectively removing the chestnut as a functioning element in America's ecosystems. It is hard to determine just what the loss of chestnut trees meant to American wildlife, but as their seeds are large and nutritious the impact was potentially catastrophic.

In 1908 an insect, the balsam woolly adelgid, was brought over accidentally from Europe. It destroyed an entire ecosystem—the forests of Frazer firs and red spruces that once graced a few high mountains in the southern Appalachians. In 1930 Dutch elm disease—another fungus—was accidentally imported to the New World, and it removed yet another important canopy tree from the continent's deciduous forests. In 1951 it was followed by another adelgid—the hemlock woolly— originally from Asia. It is presently busy blighting hemlock in the northeastern United States. Even more recently Eurasian funguses have begun to destroy butternuts and dogwoods.

Today over 300 species of alien pests and diseases are attacking North America's forests which, it seems, have not been under such threat since the asteroid impact of 65 million years ago. Have such events occurred during previous periods of enhanced migration from Eurasia? I suspect not, for in former times diseases would not have arrived without their hosts. Thus even if Asian chestnuts, carrying their killer funguses had arrived via a period of climatic warming millions of years ago, the event would not have left a continent bereft of chestnuts. Instead the Asian trees would have replaced their New World cousins and Eurasian chestnuts would have fed the continent's denizens. Depending upon the similarity of their pollen, the event might not even have been discernible in the fossil record.

TWENTY-THREE

THE FATAL IMPACT

In the 1820s Fanny Trollope, that perceptive, sharp-tongued traveller, described North America as 'a vast continent, by far the greater part of which is still in the state in which nature left it, and a busy, bustling, industrious population, hacking and hewing their way through it'.[1] That hewing, hacking and shooting was to cause a lot of environmental damage.

The extermination of American species at the hands of Europeans began, and with luck will end, in the north Atlantic. Its first victim was the great auk or garefowl. The original penguin (though not related to the southern hemisphere birds of that name), the great auk was a flightless black and white bird that stood taller than your knee. Its breeding colonies had been pillaged since prehistory but it was the French explorer Jacques Cartier, during his 1534 exploration of the Gulf of St Lawrence, who discovered its last stronghold—a vast, hidden breeding ground on Funk Island.[2] In the centuries that followed, the huge aggregations that came to nest on the island were ravaged by one boatload of visitors after another. The birds fuelled European explorations, baited cod-hooks and, in their dying days, supported taxidermists eager to turn a profit from selling stuffed and mounted specimens. The last two documented individuals were killed in Iceland and sold to a taxidermist in June 1844.

One hundred and fifty-four years later on 31 July 1998, the journal *Science* published an article sternly warning the world that a large member of the stingray family, the barn-door skate (*Raja radiata*), could become the first marine fish to be made extinct by humanity. The skate, whose size is reflected in its common name, lives only on the Grand Banks and adjacent areas of New England. Forty-five years ago fishermen working its territory on St Pierre Bank hauled it up in about 10 per cent of their trawls, yet not a single creature has been seen there in over twenty years. This is hardly surprising, for in the last years before the the Atlantic cod fishery closed every area of the banks was trawled on average every four months. Skates breed slowly and have few young and the cod nets appear to have caught every barn-door skate on the banks. They were, incidentally, a mere by-catch in the industry—hauling them up was a waste of time. Today the last few barn-door skates hold out in water a kilometre deep off Newfoundland. This area has been recently designated to support a new fishery—based on Greenland halibut. Unless something is done very soon the gigantic barn-door skate will certainly join the original penguin as another victim of the great North American expansion.[3]

In between these oceanic bookends of fish and fowl lies a history of ruthless environmental exploitation, the audacity and imbecility of which leaves one gasping for breath. In the words of Henry Fairfield Osborn, president of the New York Zoological Society in the early twentieth century, 'nowhere is Nature being destroyed so rapidly as in the United States...an earthly paradise is being turned into an earthly Hades; and it is not savages nor primitive men who are doing this, but men and women who boast of their civilisation'. It was a cry heard over and over throughout North America, yet for a century it seemed as if nothing could stop the slaughter. This is the sad story of the economic machine that ate the life of a continent, and it was not just animals that were fed into its maw, but people and cultures too.[4]

When the Pilgrim Fathers stepped ashore at New Plymouth they imagined that they were entering a new and virginal land, yet in 1621 New England was, some writers have said, more like a widow than a virgin. Three years earlier her native peoples had been visited by a terrible pestilence, possibly the plague or smallpox. The Patuxet plague ensured that for some years there would be no real competition for land between new colonists and old. In such circumstances the greatest benefit to either group was to be had from honourable trade. The Indians benefited from the trade goods brought by the Europeans while, at least at first, the Europeans were dependent upon the Indians for their survival.

There was a hidden cost in the trading business—a trade in germs— and it was one that favoured the Europeans, for continuing outbreaks of Eurasian diseases among the Indians emptied more and more land into the hands of the expanding European population. The manner in which this occurred was incredibly fortuitous for the newcomers.

In 1634 a violent epidemic of smallpox broke out among the Pequots living inland along the Connecticut River. 'It pleased God,' William Bradford wrote, 'to visit these Indians with a great sickness and such a mortality that of a thousand, above nine and a half hundred of them died, and many of them did rot above the ground for want of burial.' Bradford goes on to describe that

> for want of bedding and linen and other things they fall into a lamentable condition as they lie on their hard mats, the pox breaking and mattering and running one into another, their skin cleaving by reason thereof to the mats they lie on. When they turn them, a whole side will flay off at once as it were, and they will be all of a gore blood,

most fearful to behold. And then being very sore, what with cold and other distempers, they die like rotten sheep.[5]

The utter dependence of many Europeans on the Indians is revealed by Bradford's mention that 'the three or four Dutchmen' living among the Indians (and almost certainly the vector for the disease) 'almost starved before they could get away'. The terrible affliction soon spread to other tribes. To the credit of the English, 'though at first they were afraid of the infection' they went to help the Indians 'and daily fetched them wood and water and made them fires, got them victuals whilst they lived; and buried them when they died'.[6]

The Mandan tribe, which lived along the upper Missouri River, was blighted by smallpox in the fall of 1837. The disease was carried to Fort Union in the body of one Jacob Halsey, a passenger aboard an American Fur Company steamboat. Halsey had been vaccinated and quickly recovered, but the Indians were not so fortunate. Although the European fur traders attempted to impose quarantine, they did not send the steamboat back. Within the course of two months this fatal decision was, according to one observer, to cost the lives of 1569 out of the 1600 Mandan. The thirty-one survivors were made slaves by the Riccarees, who settled on the newly vacated Mandan land, much as the pilgrims had done in Wampanoag territory a century and a half earlier. The last Mandans died pathetically, for when the Riccarees were attacked by the Sioux the Mandans ran through a defensive picket onto the prairie, calling out to the Sioux to kill them because 'they were Riccaree dogs, that their friends were all dead, and that they did not wish to live'.[7]

Tragically, the smallpox epidemic was to spread well beyond the Mandan, for during 1838 it would kill ten out of every twelve Assiniboin in the region of Fort Union, as well as about 6000 Blackfeet. Of the Assiniboin who survived, about half had been vaccinated by fur traders during visits to the Hudson's Bay Company's trading posts in Canadian territories. The Hudson's Bay Company had been trading with

Indians for nearly 200 years by the time of the epidemic, and well knew the importance of a healthy Indian population to its business. It sent supplies of smallpox vaccine to its traders for the benefit of their Indian trading partners. Perhaps predictably, the Americans failed to provide any such protection to the Indians on their frontier.[8]

Guns, Germs and Steel by Jared Diamond lists in its title the three factors that permitted Eurasian supremacy in the Americas, Australia and other colonial arenas. But which—the guns, the germs or the steel—was the most important? The answer, Diamond has no doubt, is germs. Although estimates vary widely, by 1492 the human population of the Americas may have reached 57 million, of whom 21 million lived in Mesoamerica. Eighty years later it had shrunk to 18 million. Not even the most bloodthirsty conquistadors could have effected such carnage, for among the deadly trio only germs had the ability to kill on a scale like this. Indeed it is thought that disease had wiped out 90 per cent of the people living in Mesoamerica, Peru and the Caribbean by 1568.[9]

For a long time it was believed that the Indians gifted at least one disease, syphilis, to the invading Europeans; and a most virulent pox it was in its early stages, turning penises and noses black and rotten and often killing its victims within weeks. After 1493 it raged through Europe, altering the course of dynasties and perhaps changing European morals before it assumed a less virulent form. New archaeological evidence, however, suggests that syphilis was long resident in Eurasia, for signs of treponemal (syphilitic) infection have been discovered in 1000-year-old human bones from southern Africa, and in 2500-year-old Greek skeletons from Italy. Just why syphilis became so deadly after 1493 remains unclear, but it seems probable that it had something to do with the arrival of a New World strain carried by Columbus's crews.[10] It is possible that when the New and Old World varieties came into contact, a more deadly manifestation of the disease was spawned. Why this happened in Eurasia and not the New World, however, remains unexplained.

The extent to which disease assisted the European conquest of North America can be gauged by comparing experiences of European colonisation with those of Afro-Eurasia. European colonists in Africa and Asia were every bit as brutal as their counterparts in the Americas, but they almost never managed to extirpate the indigenous populations, who were often as poorly armed as the American Indians. As Diamond points out, Afro-Eurasia forms one vast cesspool of verminous microbes, for its human populations have been sharing diseases since time immemorial. The Americas, however, were microbially naive and the European victory there is testimony to the power of germs over guns and steel in shaping world history.

Today attitudes to Indian health on the continent have changed and it is now a major concern to health professionals. Yet for all that modern medicine can offer, Indian death rates remain significantly higher than those of European Americans. As late as the 1960s the average age at death for Indians in North America was forty-three, and 500 out of every 1700 Indian infants died in their first year of 'preventable diseases'.[11]

The American frontier has a reputation for extreme violence against Indians, and one wonders if justice ever prevailed, and whether the hands that bore the guns and steel sometimes offered kindness. Surprisingly, even some Indian-hating Puritans showed themselves capable of compassion and justice towards those 'cohorts of the devil' and tended to Indians stricken with disease. On visiting the small Indian settlement of Cummaquid (now Barnstable, Massachusetts), one Puritan wrote:

> One thing was very grievous unto us at this place. There was an old woman, whom we judged to be no less than a hundred years old, which came to see us because she never saw English, yet could not behold us without breaking forth into great passion, weeping and crying excessively. We demanding the reason of it, they told us she had three sons who, when Master Hunt was in these parts, went aboard his ship to trade with him, and carried them captives into

Spain...by which means she was deprived of the comfort of her children in her old age. We told them we were sorry that any English-men should give them that offence, that Hunt was a bad man, and that all the English that heard of it condemned him for the same; but for us, we would not offer them any such injury though it would gain us all the skins in the country. So we gave her some small trifles, which somewhat appeased her.[12]

As early as 1638, two Englishmen were executed for murdering an Indian, yet such judgments became rare as the frontier moved west and the balance of power between European and Indian more one-sided. By the time of Pontiac's war of 1763–64, unspeakable atrocities were being perpetrated on the Indians by 'civilised' men. Lord Jeffrey Amherst, commander-in-chief of the English army, was perhaps the most fiendish, for it was he who began germ warfare in North America. When the English commander at Fort Pitt invited the Delaware Indians to negoti-ate a peace settlement under a flag of truce, Amherst ordered that they be given smallpox-infected blankets from the fort hospital. The result-ing deaths of men, women and children were catastrophic.[13]

After the War of Independence matters only got worse for the Indians. Most fought on the side of the British because its government offered them future protection from frontier violence and because it intended to limit European settlement to the east of the Appalachians. The land hunger of young America, however, was insatiable. Americans were liter-ally willing to wade through Indian blood in order to take land. Settled Indians were sometimes befriended by whites, who had themselves written into the red men's wills before killing their 'friends' in cold blood. Tragically, straightforward murder was even more common, for again and again whole groups of Indians were rounded up and killed simply for their land.

The white land hunger was indissolubly linked with patriotism in a doctrine of manifest destiny—a sense that God and fate had destined them to inherit the New World in its entirety. This supreme expression of self-confidence had been fostered by the effect of germs on the Indians, by the ecological and social release experienced by the Americans and by the vast bounty reaped on their soil frontier. It was such a heady mix that for a while the Americans imagined that they were God's chosen people—a people perhaps who could do wrong on their frontier.

As the European frontier advanced, the Indians acquired all manner of European material goods, ideas and institutions along with the diseases that decimated them. Some of these acquisitions were to transform Indian cultures and to provide last-minute defences. An outstanding example of cultural change resulting from diffusion of European influence concerns the last tribes to maintain their autonomy—the plains Indians. When first contacted by the Spaniard Coronado in 1541, the Apaches were living in small family groups, hunting and travelling on foot and using dogs rather than horses to carry their limited possessions on a *travois* (a hide slung between two poles). In the *travois* we see an echo of the Arctic ancestry of these Athabaskan people and their boreal relatives' dog sleds. The Apache economy was based entirely on bison. They used bison skins for their housing, clothing and ropes, bison bones for tools, dung for firewood, bladders for jugs and meat for food. Killing bison on foot must have been hard work and the eventual acquisition of horses and guns by the plains Indians led to a lifestyle revolution.

So profound was the change brought by horses that by the 1830s the artist George Catlin wrote, 'A Comanche on his feet is out of his element...almost as awkward as a monkey on the ground...but the moment he lays hand upon his horse, his face even becomes handsome, and he gracefully flies away like a different being.'[14] Although their economy remained centred on bison, the plains Indians could now forage over larger areas and hunt more effectively. This led to a population explosion as well as to the development of larger groups. Horses and guns, along with the demographic changes they wrought, gave the plains tribes resilience in the face of European oppression that was unmatched in other Indian groups. It was a change much resented by the homesteaders, for as Frederick Jackson Turner noted, 'long before the pioneer farmer appeared on the scene, primitive Indian life had passed away. The farmers met Indians armed with guns.'[15]

Few Indian groups adopted European lifestyles as wholeheartedly as the Cherokee, who lived in what is now Georgia. The Cherokee genius Sequoyah, after whom the stately redwoods are named, developed a written form of the Cherokee language. Soon they were publishing newspapers in their own languages and had opened their own schools. Some Cherokee became highly successful entrepreneurs. Lewis Ross, for example, lived in an elegant white plantation house and owned a mill, stores, ferry-boats and over forty black slaves.[16]

Despite their adoption of western culture the fate of the Cherokee was particularly dismal. In 1830, prompted by the perfidious President Andrew Jackson, whose life had once been saved by a Cherokee, Congress passed the Indian Removal Act. The Cherokee fought the Act in the Supreme Court and, in 1832, Chief Justice John Marshall found in their favour. The whites, encouraged by Jackson, chose to ignore the highest judicial power in their land and began the process of dispossessing the Indians.

Many Americans were horrified by what followed. A New York newspaper agonised:

> We know of no subject, at the present moment, of more importance to the character of our country for justice and integrity than that which relates to the Indian tribes in Georgia and Alabama, and particularly the Cherokees in the former state. The Act passed by Congress, just at the end of session, co-operating with the tyranni-cal and iniquitous statute of Georgia, strikes a formidable blow at the reputation of the United States, in respect to their faith, pledged

in almost innumerable instances, in the most solemn treaties and compacts.[17]

Within a year the fine Cherokee mansions, schools and businesses were surrounded by 'white men...like vultures...watching, ready to pounce upon their prey and strip them of everything'. Ahead lay the 'trail of tears', as the Cherokees' trek westward to their appointed reservation in 'Indian Territory' became known. They were forced to march at the most dangerous time of year, and between 4000 and 8000 died of starvation, exposure and despair. Their suffering did not stop there, however, and they were subject to subsequent relocation as their 'Indian Territory' repeatedly turned into white man's land.

Guns, germs and steel were not always the most potent weapons in appropriating land—the pen had its uses too. By 1871, when the North Americans decided that the 'treaty' was no longer a useful weapon in their Indian wars, the United States had made more than 370 individual treaties with various Indian groups, every single one of which had been violated by European Americans.[18]

The final phases of the Indian war were just as shameful and unrelenting. In the twenty-five years between 1865 and 1890 the United States Army alone killed 6000 Indian men, women and children. It was an expensive business, for by 1870 the campaign was costing approximately one million dollars for every dead Indian. Such a high ratio of effort for return has not been matched in any other North American extermination program, except perhaps the campaign against the last wolves in the US.[19]

Guns and steel had their last substantial use in the Indian wars three days after Christmas Day 1890, at a place known as Wounded Knee Creek, South Dakota. There, about 300 unarmed Sioux men, women and children were murdered by the Seventh Cavalry, George Custer's unit, which had been shamed at Little Bighorn in 1876. In their frenzy

for revenge they even cut down thirteen of their own men in the cross-
fire. Several of these gallants received congressional medals of honour
for their heroic deeds in action.[20] As a codicil to this tale, two more
Indians were to die under European fire at Wounded Knee—this time
in 1973 when Indians protesting at the site were surrounded by 300
national guardsmen and US marshals who opened fire on them. Progress
in race relations is evident, however, for this time no medals for heroism
were issued to the men holding the guns.

Even the massacre at Wounded Knee Creek did not extinguish
autonomous Indian America. In California a tiny band of southern Yahi
Indians had escaped the numerous massacres that blight that state's
history and found refuge in the hills near Oroville, north of Sacramento.
In 1911 the last survivor of this group made contact with the Europeans.
The man was 'emaciated to starvation', and when found was cringing in
a corral at a rural slaughterhouse. His name was Ishi, and for the next
four and a half years of his life he occupied a room in a museum in San
Francisco. Ishi's death in 1916, at a time when Europeans were slaugh-
tering each other in a frenzy of mustard gas, cannonades and barbed
wire, marks the end of 'savage' America.

TWENTY-FOUR

AMERICA UNDER THE GUN

As the reach of the United States spread west, much of the native fauna and flora of the continent came to be seen either as a resource to be exploited to the full, or as a pest to be gotten rid of. In the process, men blind to nature would blast marvels from the face of the Earth, destroying forever the best of America's wildlife. If one creature is emblematic of the vitality, fecundity and sheer exuberance that was North America at this time it is the passenger pigeon, for it was a species that seemed to have lived life with the same energy as the young nation itself. In a sense it inhabited its own self-renewing frontier; nomadic flocks would strip an area before moving on, allowing the forest to recover before they came again.

Two centuries ago the passenger pigeon was the most abundant bird on the planet. Perhaps four North American birds out of every ten belonged to this single species. So huge were their numbers that as the great flocks passed overhead their droppings would fall like snow, leaving a whitened landscape in their wake. These gregarious birds were denizens of the great deciduous forests of the eastern half of the continent, and there they pursued an extravagant lifestyle that only a continent as rich in resources as North America could afford. Their every activity, it seems, was performed at an octane-charged rate.

The sleek, swift birds sped through the forests at nearly 100 kilometres per hour; the ornithologist John Audubon wrote somewhat prophetically 'when an individual is seen gliding through the woods and close to the observer, it passes like a thought, and on trying to see it again, the eye searches in vain; the bird is gone'. To fuel their flight they would eat almost anything that would fit down their gullet—acorns, nuts and caterpillars were favourites—but if they found a tastier morsel they would eject from their crop whatever was already there to make room for the new. Pioneers wrote of flocks wheeling as they fed like a giant threshing machine, stripping the ground before them as they went, the circling action giving each bird a fair chance to feed. Their breeding was also frenzied; the single egg hatched after just twelve or thirteen days and the squab was fed by the parents for just two weeks. The adults then left and the young was on its own. It scrambled to the ground, fending for itself as best it could and, if it survived, was flying and feeding independently thirty days after the egg had been laid.

What is remembered most about the birds is their breeding aggregations—some flocks were estimated to contain up to two billion individuals. Such vast congregations of nesting birds were doubtless designed to overwhelm predators. Indians would gather in their hundreds whenever a nesting site was located and eat nothing but squabs for the month or so that they were available. They were joined by wolves, bobcats, hawks and a multitude of other carnivores. Despite the best efforts of these serried ranks, enough young pigeons survived for the species to thrive for thousands of years.

The pigeons met their match, however, in the Europeans, who by the middle of the nineteenth century were devising new means to harvest resources on the grand scale. Sol Stephan, the zookeeper who looked after the last passenger pigeon at Cincinnati Zoo, noted: 'The beginning of the end began with the invention of the breech-loading shotgun about 1870. With this gun a hunter could kill 50 birds before his comrade with a muzzle-loader could prepare his charge.' Stephan somewhat pathetically added: 'Although the pigeon did no particular harm and was a very delectable dish, it became the popular thing to kill the birds.' So popular was it that pigeon-killing competitions were organised in some areas, one competitor having to produce 30,000 carcasses before taking the wreath of victory.[1]

Nature writer Peter Matthiessen perceptively called the passenger pigeon 'a biological storm'. An unforgettable account of a nesting colony was penned by John Audubon in the 1860s:

> The sun was lost to our view, yet not a single bird had arrived. Suddenly there burst forth a general cry of '*Here they come!*' The noise which they made, though yet distant, reminded me of a hard gale at sea…As the birds arrived, and passed over me, I felt a current of air that surprised me. Thousands were soon knocked down by polemen…perches gave way under the weight with a crash, and, falling to the ground, destroyed thousands of birds beneath…It was a scene of uproar and confusion. I found it quite useless to speak, or even to shout, to those persons who were nearest me. The reports, even, of the nearest guns, were seldom heard; and I knew of the firing, only by seeing the shooters reloading…the pigeons began to move off…and, at sunrise, all that were able to fly had disappeared. The howlings of the wolves now reached our ears; and the foxes, lynxes, cougars, bears, raccoons, opossums, and pole-cats, were seen sneaking off from the spot, whilst eagles and hawks, of different species, accompanied by a crowd of vultures, came to supplant them and enjoy their spoil. It was then that the authors of all this devastation began their entry amongst the dead, the dying and the mangled. The

pigeons were picked up and piled in heaps, until each had as many as he could possibly dispose of, when the hogs were let loose to feed on the remainder.[2]

European naturalists of the day found it difficult to accept the veracity of such fabulous-sounding accounts. English squire Charles Waterton, who read Audubon's narrative soon after it was published, commented dryly that either Mother Nature 'herself was in liquor, or her wooer in hallucination', to have produced such a phenomenon. But Audubon simply wrote down what he saw, for his testimony is backed up by others, and from records of the carcasses of the monumental slaughter that were shipped east to ready markets aboard the new railroads. Often-times these markets were glutted and thousands of birds were dumped.

Although laws designed to protect the species were enacted, they were ignored by the populace and within a few years mass destruction had overcome mass reproduction. By 1890 the species was in serious trouble and the last wild bird was killed in Ohio in the spring of 1900. Fourteen years later, Martha, the last passenger pigeon, looked into the now empty skies that had once thronged with her kind. She died in her cage at Cincinnati Zoo on 1 September 1914 and her stuffed skin now forms a pathetic display in the National Museum of Natural History in Washington. Many North Americans would simply not believe that it was they who were responsible for the end of this incredible creature. Instead, ridiculous rumours circulated to account for the absence of the great flocks, including one that they had all flown to Australia![3]

The years 1914–18 were bad ones for Cincinnati Zoo, for they saw out the last Carolina parakeet, as well as Martha. The beautiful parakeet— North America's only native parrot—was common enough in the nineteenth century to be counted an agricultural pest. This, plus its habit of returning to flutter around fallen comrades, was enough to ensure its end in a land where everyone carried a gun.

The demise of the passenger pigeon had unfortunate repercussions on other bird species. One birdwatcher remarked: 'When the passenger pigeon began to decrease in numbers, about 1880, the marksmen looked about for something else to take its place in the market in the spring. They found a new supply in the great quantities of plover and curlew in the Mississippi Valley at that season...They were shot largely for western markets at first; they began to come into the eastern markets in numbers about 1886.' This mad shooting spree saw the trumpeter swan and whooping crane brought within a few shotgun blasts of extinction—only a few dozen of these most magnificent birds survived by the early twentieth century. In the mad melee, even the ubiquitous herring gull was reduced in number, breeding at just a single colony on the Atlantic coast.[4]

The Eskimo curlew was once one of the most abundant shorebirds in North America; in the nineteenth century its flights were reported to have 'darkened the sun'. One migratory flock that flew over Nantucket in 1863 exhausted the island's supply of shot and gunpowder after 7000 or 8000 birds were blasted from the sky. In 1891, while searching for its breeding grounds in northern Canada, the ornithologist–explorer Roderick MacFarlane wrote that 'among the many joyous bird notes which greet one while crossing these grounds, none seemed more familiar or pleasanter than the prolonged mellow whistle of the Esquimaux Curlew'. Since the time of Captain James Cook these birds had provided 'the greatest delicacy', and long before that they had been harvested sustainably by Native Americans.[5]

The arrival of commercial shooters, who found a desirable resource in the numbers of Eskimo curlew, presaged enormous danger. Although each curlew weighed under half a kilogram, they were keenly sought because of the great amount of fat they laid down in the fall. Marketed as 'doughbirds' the creatures were just the right size to fit in a tin, and canning factories were set up in Labrador in the 1880s. By 1890 commercial exploitation had led to a rapid decline in their numbers, reducing them to the verge of extinction. Today, over a century later, the species still shows no sign of recovery. The last confirmed sighting was in 1974, and the Eskimo curlew may yet prove to be America's most recent lost bird species.

By the early twentieth century, after blasting their way through the larger species, commercial hunters were reduced to the pitiful expedient of shooting swallows, sometimes killing over a hundred of these tiny birds with one shot. Their pathetic, mouse-sized carcasses graced the markets of New York and Philadelphia in inordinate numbers because most of the larger bird species were simply no longer available.[6]

That venerable part of the European economy in North America, the fur trade, reached appalling proportions at this time as well. In a book written at the end of the nineteenth century to promote the International Fur Store in London, it was reported that each year England imported the pelts of 50,000 wolves, 30,000 bears, 22,000 American otters, 750,000 raccoons, 40,000 cats, 50,000 to 100,000 pine marten and 265,000 foxes. Most came from North America. Here was over-exploitation of the natural resource base on the truly grand scale. As a result, by the middle of the twentieth century the grizzly and wolf would be all but extinct in the contiguous United States, with only the vastness of Canada and Alaska providing a refuge.[7]

'We come now,' wrote William Hornaday in 1887, 'to a history which I would gladly leave unwritten. Its record is a disgrace to the American people in general, and the Territorial, State, and General Government in particular. It will cause succeeding generations to regard us as being possessed of the leading characteristics of the savage and the beast of prey—cruelty and greed.' Hornaday was referring to the extirpation of the American bison, particularly the most disgusting and wasteful stage of its slaughter, which occurred between 1830 and 1868.[8]

Hornaday, a veritable Boswell of the bison, records that the first European to see the creature in its native habitat was the Spanish adventurer Cabeza de Vaca. A prototypical Buffalo Bill, Cattle Cabeza, as his name translates, encountered bison in about 1530 as he wandered lost in south-western Texas, seeking a way home to Mexico. It was, however, another Spanish explorer, Francisco Coronado, who left us the most memorable early description of the 'crookebacked oxen'. When he saw them in 1542, the Texas panhandle was as full of them as 'the mountain Serena in Spain is of sheepe':

> The first time we encountered the buffalo all the horses took to flight on seeing them, for they are horrible to the sight. They have a broad and short face, eyes two palms from each other, and projecting in such a manner sideways that they can see a pursuer. Their beard is like that of goats, and so long that it drags the ground when they lower their head...They always change their hair in May, and at this season they really resemble lions...Their tail is very short, and terminates in a great tuft. When they run they carry it in the air like scorpions...Another thing which struck us was that all the old buffaloes that we killed had the left ear cloven...[9]

This extraordinary, archetypal American creature is now so unfamiliar that Coronado's quite accurate description seems fantastical. Once found over a third of the continent, from the tidewater of the Atlantic coast to the eastern fall of the Sierra Nevada, it was, wrote Hornaday, 'the most economically valuable wild animal that ever inhabited the American continent'.[10]

Bison were never accurately censused until their numbers dropped so low that they could be individually counted, but estimates of the early nineteenth-century population range between 30 and 60 million. Such vast figures can only be made sensible to us today through written accounts of this animal *en masse*. In 1795, a French trapper recorded that 7360 buffalo were drowned or mired along the Canadian Qu'Appelle during the spring break-up. Another recorded that drowned herds floated past his trading post, setting up such a prodigious stink that he was unable to eat his dinner. Billy Dixon crossed paths with a herd in 1874. One morning after breakfast, Dixon heard that 'familiar sound come rolling toward me from the Plains—a sound deep and moving, not unlike the rumbling of a distant train over a bridge. In an instant I knew what was at hand. I had often heard it. I had been listening for it for days, even weeks.' The sound was coming from bellowing bison bulls, a 'continuous, deep steady roar that seemed to reach the clouds'. The bellowing had carried for more than sixteen kilometres over the plains, and after riding out 'as far as the eye could reach, south, east and west of me there was a solid mass of buffalo—thousands upon thousands of them—slowly moving toward the north'.[11]

The story of the evolution of the bison has been told in preceding chapters, but it is important to restate its place in North American ecology. Bison dominated the vast ungulate communities of the Great Plains. These herds had assembled themselves with mind-boggling speed—only 13,000 years—and nothing like them had ever existed previously on the continent. The millions of bison roaming the plains in 1492 were critical to this new ecosystem, for the vegetational diversity of the prairie was dependent on them, which in turn benefited the lesser ungulates such as American antelope and elk, as well as many other species. There were only two significant predators in this ecosystem, the wolf and the plains Indians, both of which occurred at extraordinary densities for such large carnivores (there being an estimated 400,000 wolves in the US and 200,000 to 400,000 plains Indians). Both populations were utterly dependent upon the bison for their survival. The genus *Bison*, which had first appeared in North America south of the ice sheets just half a million years earlier, had thus elevated itself into a position of absolute ecological dominance, and by 1492 was a keystone species. Its near extinction was to have important implications for America's rangelands.[12]

Large mammals such as bison are important to ecosystems because they have a miraculous effect on soil fertility. In general soil fertility declines once rainfall reaches 800 millimetres per year because of leaching. On African savannas that still support a megafauna, however, soil nutrient levels peak in regions receiving 900 to 1100 mm of rain per year. This is because the great herbivores act as nutrient sinks, continuously doling out the precious nutrients in their urine and faeces. In effect they store nutrients, preventing them from being lost through the water. They also trample mature vegetation, stimulating new growth. These characteristics make megafauna-dominated savannas among the most productive regions on Earth.[13]

The key to the bison's role in the prairie ecosystem lay in the fact that the great grasslands were piss-driven. Buffalo urine was the critical fertiliser for the grasses, and their fertility in turn dictated where bison grazed. It was a self-reinforcing cycle that, when coupled with the vast migrations of the herd, kept the prairie an exceptionally productive place. Even bison wallows were important, for distinctive plant communities grew by them that were favoured by other herbivores such as American antelope. The whole ecosystem had in effect organised itself around the bison's survival, and it was an ecosystem of enormous proportions, for the bison were accompanied by around nine million American antelope, 3.6 million deer, two million elk and two million bighorn sheep.[14]

Hornaday summarised the cause of the buffalo's destruction as 'the descent of civilisation'. 'Civilisation' was a word Hornaday uttered between clenched teeth, for the story he documented was informed by his enormous shame and contempt for the actions of his countrymen. From the very beginning, he lamented, the slaughter involved unforgivable waste. In the very early days most bison were shot for their skins, which were made into blankets or robes. Later they were shot for their tongues or from railway carriages for sport. According to Hornaday these sportsmen were 'murderers'. The period of systematic slaughter proceeded in a 'business-like, wholesale way'. Hornaday wrote that 'perhaps the most gigantic task ever undertaken on this continent in the line of game slaughter was the extermination of the bison in the great pasture region by the hide hunters. Probably the brilliant rapidity and success with which that lofty undertaking was accomplished was a matter of surprise even to those who participated in it.'[15]

A decisive blow against the bison was struck by chemists, who in 1871 perfected a tanning process for buffalo hide. The British Army added to their perfidy by proclaiming that buffalo leather made the very best military footwear. From that moment no bison was safe. Between 1871 and 1874 'little else was done' in the regions of Dodge City, Wichita and Leavenworth 'except buffalo killing'. The hundreds of thousands of skins that Colonel Dodge saw sent to market were scarcely an indication of the extent of the slaughter. For each hide sent, five went to waste. Many animals were shot and then allowed to wander off to die, while others were skinned so inexpertly that the hides had to be discarded. Hornaday laments that the curers were so ignorant of their trade that half of the skins they obtained rotted.[16]

Eighteen eighty-four was the year the buffalo lost the battle for the plains. Theodore Roosevelt recalled meeting a rancher who had travelled a thousand miles that year and who related that he was 'never out of sight of a dead buffalo, and never in sight of a live one'. The year before 40,000 buffalo hides had been shipped east, but in 1884 only 300 were sent. The days of the bison—and the bison hunter—were over.[17]

On 1 May 1889, as he was putting the finishing touches to his *magnum opus*, Hornaday wrote that 'the nearer the species approaches to complete extermination, the more eagerly are the wretched fugitives pursued to the death whenever found. Western hunters are striving for the honor (?) of killing the last buffalo.' Were it not for the vigilance of park rangers, the 200 buffalo that found refuge in Yellowstone National Park 'would have been shot years ago by Vic. Smith, the Rea Brothers' and others, Hornaday sighed.[18]

In all just 600 buffalo survived the slaughter. These had found refuge in out-of-the-way or protected places in Canada and the US, and herds were eventually brought together to act as nuclei for a new breeding program. The Custer State Reserve in South Dakota's Black Hills played a crucial role in this process. Around 1000 of the great beasts can still be seen trudging across the grassy valleys there, the bulls uttering their peculiar lion-like roar just as they did when Billy Dixon heard them in Texas a century before. For all their magnificence there is something heartbreaking about visiting the place. To have to come so far to see a creature that was once unavoidable on the Great Plains seems wrong, and to make it worse the buffalo in Custer are carefully managed. They are all branded, and any over ten years old are culled. This rigorous strategy has maintained a virile stock that has brought the species back from the brink of extinction, for the Custer herd has contributed greatly to the 250,000-strong buffalo herd on private lands across the nation. Yet despite their muscular bodies, the mop-headed creatures lack the wildness and majesty one expects in an American buffalo.

The demise of the bison had a profound impact on both wolves and plains Indians. The decline of the wolf is a story also written in hides. In 1880, 7000 wolf skins had been sold in the US, but by 1885, the year after the last great bison massacre, just 273 skins reached market. By 1965, the year of their nadir, just 500 wolves remained in the contiguous states of the US and there is little doubt that had Americans possessed the resolve to continue their persecution, there would be no wolves left there today. In the case of the plains Indians, their discomfort seems to have been one of the principal reasons that the US government permitted the destruction of the buffalo in the first place. The massacre at Wounded Knee Creek that heralded the end of Indian autonomy occurred in 1890, just six years after the demise of the bison.[19]

So abundant was the buffalo that even after its extermination its humble remains formed a valuable resource. The bones scattering the prairie were exploited on a grand scale during the economically disastrous closing years of the nineteenth century, when an industry grew up around collecting and grinding them for fertiliser or for use in refining sugar. What is even more remarkable is that ancient Indian bison kill-sites were also mined for their bones, and the volume recovered was sometimes stupendous. The Highwood site in the Missouri Valley, for example, yielded well over 6000 tonnes of fertiliser and bonemeal derived from buffalo remains. The skeleton of a bison typically comprises 10 per cent of its total body weight. If we allow a modest average weight of 300 kilograms for calves, cows and bulls killed at the site, then no less than 20,000 creatures met their death at the Highwood jump over some thousands of years.

The Frost Fertiliser Company of Montana, ' "The Pioneer" producer of buffalo compost', advertised a product made from bones, which was not only used for fertiliser but was fed directly to cattle! A typical site exploited by the company, known as the Taft Hill Jump in Montana, produced 150 tonnes of bison carcass and bonemeal, which was pulverised and shipped to the west coast. The site though was far from exhausted, for workers stopped only when the bone-to-soil ratio made their work too tedious. Although clearly less than 10,000 years old, the age of the deposits exploited by such companies is uncertain. Some contained whole mummified or frozen buffalo, an invaluable scientific resource that was ground up and used as manure.[20]

In terms of kilograms of matter belonging to one species, the great bison herds of the American prairie formed the greatest aggregation of living things ever recorded. Not Africa, nor Asia, nor even the seals and penguins of the Antarctic, could offer a sight to equal the sheer biomass of these vast herds. The passenger pigeon, likewise, was a wonder of the world. What would it mean to people today to be able to stand in awe, watching the great herds and vast flocks that thronged the continent just over a century ago? Were they still here, North America would rival Africa as a safari destination—a continent of marvels and wonders.

The North American species that fared worst in the late nineteenth century—the passenger pigeon, the buffalo and Eskimo curlew—were those that congregated in vast flocks or herds in order to overwhelm predators. This appears to be a unique aspect of the North American extinctions, for other places that have suffered extinctions, such as Australia and the oceanic islands, lost the rarer, larger or more solitary species, rather than those that congregated *en masse.*

The herding behaviour of the bison, and possibly the passenger pigeon, first arose as an adaptation to avoid the human predator, and this strategy protected these species for at least 13,000 years. During the nineteenth century, however, the European Americans were developing a new economy. It was an economy based on systems of mass production and mass exploitation that needed enormous resources to operate. The key decade was 1880–90, for only then did European machinery become sophisticated enough to destroy the great herding and flocking species. Their enormous numbers provided an ideal teething rusk for an economic machine that would soon produce cars in the millions and hamburgers in the billions.

During the course of the nineteenth century North America's great herds of hoofed animals declined from an estimated 60 million to just one or two million head. People came to replace ungulates in nearly equal proportion, for over the same period the immigrant population of the US rose from five million to 75 million.[21] Today, the chicken has likewise replaced the passenger pigeon as America's most populous bird. Within a century of the demise of the bison the continent's waterways, soils and fisheries would be consumed by the machinery of this astonishing new economy, as would species and resources that lay well beyond the bounds of the nations of North America. We must now turn our attention to how this terrible machine was forged.

Twenty-five

THE MAKING OF A GIANT

At the dawn of the nineteenth century the United States of America with its population of five million had just five millionaires. The country was still a minor world power, about the size of a small European nation. One hundred years later, however, the US would support a population greater than any western European power and possess an unequalled concentration of wealth. By 1908 it had given birth to the world's first billionaire, John D. Rockefeller. Historian Daniel Boorstin saw the continent as shaping this affluent American society, even to the point of influencing her political system. 'The genius of American democracy,' Boorstin wrote in 1953, 'comes not from any special virtue of the American people but from the unprecedented opportunities of this continent.'[1]

Three factors led to this vital transformation: the construction of a transportation network that would link goods and markets, the development of the American manufacturing industry, and the creation of a nation of consumers. Americans created each of these in their own inimitable way. The completion of the Erie Canal in 1825 was a keystone achievement. Earlier in the history of the continent a great inland water-way—the Bearpaw Sea—had divided North America in two. Now an inland waterway would unite it, giving rise to astronomical growth in eastern cities like New York and in the mid-west as these regions were linked for the first time.

Prior to the construction of the canal, the Appalachian Mountains had divided European America. Farmers living west of the mountains were capable of growing vast amounts of produce which they could ship cheaply down the Mississippi by barge or raft, but how were they to receive goods in return? Shipping goods against the current of the continent's largest river was not practicable, for no one could row against it for thousands of miles. It was possible to cart goods over the Appalachian watershed but the cost was enormous. In 1817, for example, it cost nearly twenty cents per ton for every mile such goods were carted. As Congressman Porter of western New York put it in 1802, 'The great evil, and it is a serious one indeed, under which the inhabitants of the western country labour, arises from the want of a market. There is no place where the great staple articles for the use of civilised man can be produced in greater abundance or with greater ease, yet as respects most of the luxuries and many of the conveniences of life the people are poor.' The advent of steamboats helped a little, but it was not until the completion of the Erie Canal that the cost of transportation was substantially reduced. By 1857 it was less than one cent per ton per mile, a massive 97 per cent reduction on pre-canal costs.[2]

The first transcontinental railroad was completed in 1869 and twenty-one years later 150,000 miles of track had been laid in the United States. By then much of the continent had been converted into a quarry and a farm, linked to cities by this cast-iron web. Here was a network for creating land speculation on a scale hitherto unimaginable, and with the railways came a land boom that made the wagonloads of earlier homesteaders look trivial in comparison.

The American Civil War in many ways represents *the* decisive moment in the country's history, for it definitively resolved the tensions that had been evident in British-settled America since its inception. It was a struggle against forces that had the potential to convert North America into another Europe—a constellation of rival states descending into centuries of self-destructive feuding. If resisted, however, the US might remain one nation. Abraham Lincoln saw the situation clearly—the war was fought principally to maintain the union.

The guns that won the war also transformed American society, not through firepower but through manufacture, for one of the most influential developments of nineteenth-century America was a transformation in the processing of materials, particularly in factories. These changes would lead to the system of mass production. Henry Ford claimed to have got the idea for the assembly line from observing the great 'disassembly lines', the meat-packing works (fed by the beasts that replaced the bison) in Chicago and Cincinnati.[3] If true, then the solutions Americans devised to exploit the natural wealth of the frontier were to have direct consequences for the entire world.

The idea of producing guns with interchangeable parts was, along with so many other far-reaching ideas, first conceived in Enlightenment France. General Jean-Baptiste Gribeauval noticed that large quantities of firearms were damaged in battle, with some having broken stocks and others damaged locks or barrels. In a flash of inspiration, he saw that an army that could construct new guns from the undamaged parts would have an enormous battlefield advantage. From this he conceived a plan to standardise the armaments of the entire French Army, allowing parts in a gun to be interchanged as readily as a soldier can switch positions. Between 1786 and 1789, the year of the French Revolution, another Frenchman, Honoré Blanc, took a great step towards achieving this. In the armoury at Vincennes, Blanc produced muskets with interchangeable locks. While ambassador in Paris, Thomas Jefferson became aware of Blanc's breakthrough. Upon returning home he urged Americans to adopt the 'system Gribeauval' as it became known. After being adopted as a central tenet of military training at West Point, it quickly became entrenched as a *desideratum* in US military thinking.

David Hounshell, the historian of American manufacturing, wrote that, 'The development of the American system of interchangeable parts manufacture must be understood above all as the result of a decision made by the United States War Department...to have this kind of small arms, whatever the cost.' The cost proved to be enormous, for it would take forty to fifty years of unrelenting effort and hundreds of millions of dollars in subsidies to achieve the goal. After all that effort, American armouries would indeed turn out guns with interchangeable parts, but they would be more expensive than hand-tooled guns and of inferior quality. If market-place competition had ruled the day, the world would not have had guns with interchangeable parts.

The War Department's real contribution to the growing nation was a totally unanticipated side effect, the creation of a pool of engineers skilled in the use of machines and cognisant of a method of production that linked such machines in series to complete a given task. They were also imbued with a notion of the inherent worth of the idea of the interchangeability of parts, even in the face of the huge costs and difficulties involved. Ford's great success in developing the system of mass production drew on these engineers' expertise. It also inspired many Americans to imitation. In industry after industry—from vacuum cleaners to radios—goods were soon being assembled on conveyor belts. In Ford's success many saw 'the ideal of renewing America through mass production'.[4] What is absent, however, is any understanding that the ideology behind mass production meant a rapid depletion of the natural resources of the continent.

In later life Henry Ford became fixated with the idea that mass production could be applied to agriculture, and he became particularly obsessed with soybeans. Ford encouraged farmers in Michigan to plant them by the million, promising to find some use for them in his factories. He employed engineers to this end, and eventually operated part of his massive River Rouge factory with soybean oil. He even found uses for soybean-based fibres and plastics, but none of them ever proved economically viable. The height of Ford's vegetable mania came at the Ford Company dinner at the Century of Progress Exhibition in Chicago. The menu consisted of 'tomato juice with soybean sauce, celery stuffed with soybean cheese, puree of soybean, soybean croquettes with green soybeans, soybean bread and butter, apple pie with soybean sauce, soybean coffee and soybean cookies'.[5]

In truth, mass production had already begun to be applied to agriculture even before Ford's initiatives. By the late nineteenth century, a system known as 'bonanza farming' had become immensely popular in America. It was, according to one observer, 'the army system applied to agriculture', a broad-acre enterprise run with managers and machines. It was, in fact, akin to factory work and the first step in the food factory chain. Today such agriculture dominates in places like North America and Australia.[6]

Along with these changes to the means of production came a vast wave of immigration into America, which historian Bernard Bailyn has characterised as 'one of the greatest events in recorded history'. Beginning in 1500, over 50 million Europeans, Africans and others would flood into the new continent. In the late eighteenth century most non-slave immigrants settled in the port towns of southern New England, where they would transform the Puritan-based societies. The influx of African slaves into the south at this time was also considerable: 84,500 arrived in the fifteen years to 1775 alone. During the nineteenth and twentieth centuries the human tide sweeping across the Atlantic reached stupendous proportions and, on reaching the Atlantic shore, pushed overland, across the Appalachians and Great Plains.

A vast, mobile population was thus created in North America, its main thrust pointed ever westward. In time it would swell the population of North America north of Mexico to around 300 million, a figure almost a hundred times greater than that of pre-Columbian times.[7]

Although Henry Ford would not live to see its full effects, his methods of mass production would be applied to agriculture with the kind of monomaniacal passion of which he would have been proud. It would result in North America becoming Earth's bread basket, the Great Plains alone producing two thirds of the world's wheat. The continent would also become an exporter of a vast number of other agricultural commodities. Despite these achievements the result may prove to be the greatest disaster for the planet, and for North America in particular, in the last 65 million years of its history. These activities unleashed a monster that would drink many of her rivers dry, consume her plains, blight her deserts and sterilise her seas. Bison and passenger pigeon were a mere first course in the revels of this juggernaut. The true devastation of the continent was yet to come.[8]

The ivory-bill woodpecker, the largest of its kind, was in decline for most of the nineteenth century but it did not succumb until the twentieth. It inhabited, Audubon recalled, 'those deep morasses, overshadowed by millions of gigantic dark cypresses, spreading their sturdy moss-covered branches, as if to admonish intruding man to pause and reflect...' Unfortunately for the bird, most men were not inspired to reflection by the cypress swamps. Instead they saw valuable timber. Railroads opened the lumber frontier and, in 1890, eight billion board feet of timber was cut in the Great Lakes states alone. Within the range of the ivory-bill, the hunger for lumber (itself an American word) seemed insatiable.[9]

If a pair of ivory-bills was to successfully raise young they needed about eight square kilometres of undisturbed forest within which to find their food. By 1939 cutting had so fragmented the forests that just two dozen breeding pairs of the woodpeckers remained. In 1946 ornithologists located a population in the Singer forest tract of northern Louisiana and begged politicians and timber companies to spare its habitat, but their pleas were ignored. Ironically, the final substantial area of ivory-bill habitat was cleared to grow Ford's beloved soybeans. By 1968 only six pairs were known to exist and today the world's largest woodpecker is extinct in the United States. A tiny population survived in Cuba, where sightings were made in the 1990s. In an interesting paradox, the economic lethargy induced by communism may have saved this remnant, though hope for this appears more and more distant, as the latest news, published in May 2000, suggests that the ivory-bill has vanished from its last Cuban refuge.[10]

Of all North America's resources, the most precious are her waterways and other freshwater reserves, for without them life cannot persist. When all the God-given, well-watered and arable land in the United States had been occupied, the minds of the great capitalists whose fortunes had been made on the frontier—the railwaymen, the land speculators and the media magnates—turned to new ways to prolong the binge of spectacular profit-taking. In the west, where water had dictated settlement patterns, a destructive new avenue of exploitation was opened, for they realised that water could be used, monopolised and abused just like any other commodity. Water could make both deserts and bank accounts bloom.

The exploitation of the water frontier is, if such a thing is possible, an even more sorry story of greed than the extermination of the buffalo. As it is told by environmentalist Marc Reisner in his magnificent book *Cadillac Desert*, the battle for the waters of the west were a continuation of the rampant property speculation of the land frontier.

The earliest and most bitter fight over water in American history took place in the Owens River Valley in northern California. The first irrigation ever practised in the valley was that of the Paiute Indians who had learned from the Spanish. The Europeans who arrived in the valley in 1860 were determined to remove this red impediment to their wealth. They trumped up some cattle-rustling charges and in the fracas that followed murdered at least 150 Paiute men, women and children, the last hundred being driven into Owens Lake to drown. The Europeans, newly settled on stolen Indian land, reopened the Indian irrigation ditches and for the next fifty years proceeded to make themselves rich.[11]

Some of the murderers were doubtless still alive when the good people of Los Angeles began plotting to appropriate the Owens Valley water for themselves. This situation would end in the dispossession of the pioneers through the quasi-legal theft of their water. In the interim, the dispute would break out into a shooting war, with the dynamiting of aqueducts and destruction of water infrastructure becoming routine. By the 1970s the valley was so degraded that alkaline soil, borne on powerful winds that raked the parched earth, was endangering the health of its few remaining residents. The dynamiting of the Los Angeles aqueducts began again, but by then the war over water had been well and truly lost.[12]

The destruction of North America's waterways is arguably the greatest blow ever struck by the European Americans at the continent's biodiversity, for it blasted the oldest and most distinctive biological element on the continent. As we saw in the first act of this continental drama, it was the rivers, lakes and ponds of North America that provided a vital redoubt for life at the time of the asteroid impact. Thus many of the continent's aquatic denizens are of venerable ancestry, some dating from the age of the dinosaurs. They are the true native Americans, having evolved on the continent, and having been shaped by its unique evolutionary forces longer than almost anything else.

The diversity of life that had built up in North America's waterways over the past 65 million years was truly astounding. In the late nineteenth century they were home to spectacular numbers of freshwater crayfish, caddis flies and myriad other aquatic insects. They also sheltered over 800 species of freshwater fish, including the ancient paddlefish and gars as well as nurturing the world's greatest diversity of freshwater snails. No group, however, had diversified more wildly in North America's aquatic habitats than the freshwater mussels. Although humble in appearance, mussels are important indicators of ecosystem health. They also lead a remarkable life. Both eggs and sperm are released into the water, and when the fertilised eggs hatch they form a larvae called glochidia. These larval mussels must find a fish and attach themselves to its gills or fins to survive. After a few weeks of parasitic life they drop to the bottom, and if they are fortunate enough to alight on a cobble bank or riffle over which clean water flows, they stand a fair chance of growing into an adult. Each mussel species can attach to only one or a few species of fish and this dependence, along with their need for clean flowing water, has made them vulnerable to human-induced disruption.

An obscure stretch of the Tennessee River in north-western Alabama holds the world record for mussel biodiversity. This was established in the early 1920s when mollusc expert A. E. Ortmann sampled the shallows at a place known as Muscle Shoals. To his astonishment he found no less than sixty-three species of mussels in the clear waters that flowed through a valley dotted with wooded islands. Nowhere else on the planet even approached this diversity. Later Ortmann returned to this New World Arcadia to resurvey it, but to his dismay found that a dam had been built. The islands, clear water and beautiful riffles were gone, along with over half of the species of mussels.[13]

Today the most diverse assemblage of mussels in North America is found around Pendleton Island in south-eastern Virginia. In 1982 over forty species of freshwater mussels lived there, but surveys undertaken in the late 1990s reveal that only thirty to thirty-five species remained. It appears that 'a potpourri of widespread pollutants, from sediments to nutrients washing off farm fields to toxic chemicals buried in the mud' had killed them.[14]

In 1990 the Nature Conservancy summarised the losses and depletions from this rich realm. They reported that four out of every ten species of North American freshwater fish were either extinct or vulnerable to extinction. Half of the continent's crayfish species were similarly affected, while nearly 70 per cent of its freshwater mussels were in danger. No terrestrial group has suffered anything like this loss; the equivalent figures for birds, mammals and reptiles hover between 15 and 18 per cent. Indeed, they compare with the all-time high figures of extinction for the continent: 80 per cent of its flora was lost at the time of the asteroid impact, and 73 per cent of its large mammal genera was lost 13,000 years ago. Of all the losses North America has suffered since the arrival of the Europeans I find the rape of its fresh waters the saddest, both because it is still happening, and because it slashes at life that has been so enduring. What right do we have to extirpate creatures that have been 70 million years in the making, that have survived blight by asteroid, invasion, global cooling and ice age? To do so is as stupid and unforgivable as tossing the contents of Fort Knox into the sea. The raw figures related above may mean little to the non-expert without knowing something of the various species that have been lost. Two that I find particularly poignant are the Tennessee River snail darter, a unique fish lost to a useless dam, and the Las Vegas frog, another American who could not win against the casinos. Both fell victim to mindless development.

The methods used to bring about this catastrophe have been as diverse as they are chilling. The United States' 2.5 million dams were probably the main culprits. So great has been their impact that the National Rivers Inventory found that by 1982 only 2 per cent of the country's streams or stream segments were free-flowing, undeveloped and with outstanding natural and cultural values. River poisoning has been another weapon. One of the most famous cases occurred in 1962 when fisheries managers from Wyoming and Utah joined forces to poison over 400 miles of the Green River. Their aim was to rid the river of creatures such as the bonytail chub, which even then was endangered, along with several other rare fish species, and to release rainbow trout in the vacant river for the pleasure of fly fishermen. The poisoning program was, if anything, too much of a success, for the toxins swirled downstream into the waters of Dinosaur National Monument, killing fish in the reserve. There was of course an uproar from conservationists, but action came too late. All that has been seen of the bonytail chub in recent years are a few aged individuals, successful reproduction having apparently ceased decades ago.[15]

Fish introductions have been another major problem in the waterways. Sports fish such as trout have been introduced outside their natural range and in many instances have driven native fish populations to extinction. Pollution has also had its impact. The infamous Cuyahoga River, which flows through Cleveland, Ohio, caught fire repeatedly between 1936 and 1969. Although not reduced to flammability, dozens of other rivers were as grossly polluted. And then there are the famous invaders such as zebra mussels, up to 7000 of which have been found clinging to the shell of a single native freshwater mussel, which they were busy smothering. One senses that the loss of biodiversity in American freshwaters has some way to go yet before the cascade ceases.

By the 1930s, North America's soils as well as its water were in deep trouble. The frontier had carried farmers deep into marginal lands at a time of above average rainfall, but by 1933 the climatic pattern had reversed—rainfall dropped. Mechanised farming and overgrazing bared the soil and the Dust Bowl began to blow thick. It blew in earnest in April 1934, when a giant black dust cloud with a tan top blew over the parched fields of eastern Colorado and western Kansas. As it moved south it engulfed settlement after settlement in total darkness, darker than any eclipse of the sun and lasting far longer. Close to 390,000 square kilometres in size, people enveloped by it thought that the end of the world was at hand and rushed to churches to pray. In its wake it left not only terror-stricken humans but a landscape littered with dead rabbits and birds, and dust drifts up to two metres deep against the sides of houses.

Throughout 1934 and the first half of 1935 dust storm after dust storm raged up and down the length of the United States. You could tell where each originated by its colour: Kansas spawned brown or black storms, Oklahoma red and Texas yellow. In a sense they did herald the end of a world—the world of endless soil and frontier production methods, for all across the west, in 756 counties and nineteen states, the very fat of the land was being lost. Immense clouds of soil blew past Capitol Hill interrupting congressmen as they discussed a Bill designed to limit overgrazing of public lands. Yet even with such palpable proof of the effect in front of them some still objected to the legislation. Eventually the rich soil of North America crossed the Atlantic on the jet stream and fertilised the fields of Europe. The result, for the Americans, was a nightmare—14 million hectares of arable land destroyed, 50 million hectares severely depleted and a further 40 million hectares made marginal for agricultural use. In all, 750,000 Americans were made destitute and forced to leave their homes.[16]

By the 1950s North Americans had eliminated about four-fifths of the continent's wildlife, cut more than half its timber, all but destroyed its native cultures, dammed most of its rivers, destroyed its most productive freshwater fisheries and depleted a good proportion of its soils. They had won a great victory in war and had created one of the most affluent and self-contented societies ever seen, yet still the pillage of their natural resources was not finished. By 1999 nearly 1200 native North American species had been placed on the official endangered list, and this is a gross underestimate, for it has been reliably estimated that 16,000 species are in grave danger of extinction on the continent.[17]

Was it really necessary to destroy nature in order to create US society as we know it today? A single instance—that of the Grand Coulee Dam on the Columbia River—illustrates the type of trade-off involved. When the Grand Coulee Dam was completed in September 1941 it destroyed the most productive salmon fishery in the world. The dam was so high and formidable that spawning fish could not cross it, and no fish ladder yet designed could help. When the US was plunged into war at Pearl Harbor in December, she was ill-prepared. She had fewer soldiers than Ford had auto-workers and insufficient modern M-1 Garland rifles to equip even a single regiment. The nation did, however, have the Grand Coulee Dam, and the two million-odd kilowatts of electricity it produced could smelt a lot of aluminium. That aluminium was used to construct 60,000 aircraft in just four years, as well as to create the plutonium used in the atomic bomb.[18]

North America's pre-eminence has come about because the resources of a rich yet middle-sized continent have been mined to provide a capital base that is the envy of the rest of the world. In a sense, the over-exploitation of the frontier was akin to going out in a blaze of glory. One wonders how many North Americans living today would trade numberless bison and passenger pigeon for their position as members of a global superpower.

What is most worrying about this dismal history is that, on the frontier, ruthless exploitation, greed and senseless environmental destruction had become an honoured tradition. All of the US's defences, both political and social, were traduced by the vast wealth and influence won by the rapers of the land. Perhaps the single most important aspect of ecological release in this regard was the breakdown of authority. Various states made laws to protect wildlife such as passenger pigeons and at various times Indians were granted respite from the reign of terror, yet frontiersmen ignored the laws that irked them, and they did so with impunity. Government departments did the same. Referring to illegal land speculation related to the building of government-funded dams, environmentalist Marc Reisner says: 'It was a case of lawlessness becoming de facto policy, and it was to become more and more commonplace.'[19]

North Americans are still struggling with this legacy of the frontier, and with the legacies of even earlier extinctions. It is now time to examine some of these efforts and their effectiveness, as well as to look far into the future to see if we can discern the shape of Americas to come.

WHAT'S HOME ON THE RANGE?

North Americans began responding to the environmental threat the frontier created almost as soon as it became evident. The US Congress established the world's first national park, and today Americans take the lead in many aspects of the world environmental movement. Despite this pro-active environmental position, several problematic aspects of the environmental response have gone unchallenged.

The very first attempts by North Americans to conserve what they saw as the finest natural splendours of their continent contained the seeds of this dilemma. Reserved lands were set aside as wild lands—lands in which people should not interfere. At the most brutal and recent level, the great national parks and reserves of North America were created out of the dispossession of Indians. Having done the killing, European Americans suffered remarkable short-term memory loss, for they regarded the emptied landscape as primeval wildernesses. They then institutionalised these new wildernesses through the creation of national parks.

The Act under which the first of these parks, Yellowstone, was proclaimed in 1872, provided for the protection of the park's many wonders 'in their natural condition'. Yet at that very moment the Indians, the single most important species in the ecology of the park, were fighting a desperate rearguard action to maintain their part in land management. Their great victory at Little Bighorn was yet four years away.[1]

Most North American conservation work is guided by a dictum of the great biologist Aldo Leopold, who acknowledged that 'if we are serious about restoring ecosystem health and ecological integrity, then we must first know what the land was like to begin with'. Today America is nothing like what it was 'to begin with'—even if we go back to Indian times. A key question, as it relates to the management of the continent's reserved lands, is the extent to which Native Americans practised conservation of their animal resources. This is not an easy question to answer, for data on the past abundance of many species are very limited, and disentangling the impact of Indian hunting from that of wolves, competition from other herbivores and fire, is a difficult task.[2]

In the popular mind there is no question that Indians were excellent conservationists. It's a perception supported by the fact that by present standards, eighteenth- and nineteenth-century America abounded with game. The fact that few species became extinct between 12,000 years ago and 1492 also argues for some sort of ecological balance being struck between the Indians and their environment. Still, as we have seen with the bison, there is considerable evidence that Indian hunting was a powerful environmental force.

If a balance ever did exist between Indian hunters and their prey it must have been a fragile one, for changing technology surely challenged it. The introduction of the bow and arrow greatly increased the efficiency of hunting, and circumstantial evidence suggests that in some areas animal numbers were so reduced that Indians began to rely more on agriculture. The introduction of the horse and gun a few millennia later had the potential for a similarly profound impact, but by this time the Indian tribes had been ravaged by smallpox, measles and a dozen other diseases. They were also being harried by the Europeans, so their numbers were declining, as was their influence on the environment.[3]

On one side of the 'Indian conservation' debate stands the biologist Charles Kay, who argues that Native Americans had no effective conservation practices, at least regarding larger mammals. Kay believes that Native Americans acted in ways that maximised their individual fitness irrespective of their impacts on the environment, and that the manner in which they harvested ungulates was the exact opposite of any predicted conservation strategy.[4] Proponents of such arguments think that the large numbers of ungulates seen by early explorers such as Lewis and Clark only existed at intertribal boundaries—no-man's-lands where hunters were afraid to enter.

The story of Indian conservation as it relates to the buffalo is probably more complex and intriguing than such views indicate. Buffalo lived in vast herds that swept over the plains like fish in the sea. When the herds were nearby, meat could be had in superabundance, but then they moved on to someone else's area, and might be away for years. Such a species represents a dilemma for anyone wanting to harvest it in a sustainable manner. How could a conservation-minded Indian be sure that all of his good work in refusing to kill cows, for instance, would not be undone by the next tribe the herds swept past? Even with all of our current advantages in global communication, technology and ecological theory, the best wildlife managers in the world cannot enforce sustainable yields in such circumstances. Attempts to regulate oceanic fisheries offer the clearest recent example of our failure in this area. As a result bluefin tuna, which in terms of their movements can be thought of as the buffalo of the oceans, are set to become as rare in our day as buffalo were in our grandparents' time.

It is a paradox that the same social structure that makes it so difficult to implement a conservation strategy for a species like buffalo also
made it almost impossible for pre-Columbian Indians to endanger the
species. This is because no Indian could kill enough buffalo to make a
dent in the herds. Thus, in its development of a highly mobile, mass-
herding social structure, the buffalo had triumphed in the race for
survival over its predators.

This idea is given support by estimates of post-Columbian Indian
and buffalo populations. If there were just 30 million buffalo (an estimate
on the low side) on the plains, then Indians could have taken 2.1 million
animals in an average year and not reduced the population. As most
estimates put the total number of plains Indians at around 200,000, and
no more than 400,000, that's a lot of buffalo per Indian per year. In order
to get a truer picture we must take into account other sources of buffalo
mortality such as wolves, disease and accident, but even so it appears
that, most of the time, there was plenty of buffalo to feed the Indian
population. So while there is no evidence that Indians ever conserved
buffalo, before the nineteenth century they had no need to do so.[5]

Whether Indians ever effectively conserved elk is less clear, but I know
of no well-documented Indian conservation practices relating to this
species. It is clear that elk numbers increased in places such as Yellowstone after Indian hunting ceased, which suggests that hunting was
having a real impact on their numbers.

Indians did, however, implement conservation practices for more
sedentary species. The best examples come from eighteenth- and
nineteenth-century accounts of beaver harvesting. The northern
Algonquians controlled the hunting of beaver in defined areas known
as family hunting territories. These territories were passed down from
father to son. Traditional practices such as always leaving a pair of
beavers in every den, or of leaving juveniles or animals whose pelt was
in poor condition alive in the den, have all been recorded in these areas.

Interestingly, even in these family hunting territories there were no restrictions on the hunting of migratory species such as caribou.[6]

A greater challenge than understanding the impact of Indian hunting awaits the managers of reserved lands in North America, for the eco-system peopled by Indian, elk and buffalo is different from the one within which most of the continent's organisms evolved. They were shaped by an earlier era, one dominated by megafauna, and today almost all of the continent's large mammals, from mammoth to lion, are gone.

I believe that at the dawn of the new millennium North Americans face a problem that has been 13,000 years in the making, but which is only now reaching crisis point. The debate over how to manage North America's reserved lands has become bogged down and devoted to fiercely contested yet trivial detail such as how many elk should be allowed in places like Yellowstone. It is not the elk themselves that are the problem but the effect they have on trees and bushes and, in partic-ular, on aspen.

Ever since the great glaciers retreated and the continent lost its megafauna some 13,000 years ago, aspen has almost never grown from seed in the valleys and plains of the Yellowstone region. All aspen growing there are clones derived from vegetative reproduction. These clones have survived for more than 10,000 years under all sorts of conditions, but for most of this century have been in decline, with researchers predict-ing that they will be extinct in Yellowstone within the next fifty years if the trend continues.[7]

The exact cause of, and even extent of, aspen decline is uncertain. Yet, even if an increasing number of elk are the sole reason for the decline, many managers would not limit elk numbers, for they are opposed to any sort of intervention in these 'wild' lands. And it is not only elk that can be fatal to aspen. All across the northern rim of the American prairie, shimmering-leaved aspen grow in parkland with a grassy under-storey. This beautiful environment is new—the indirect result of European impact. Here aspen is expanding, so the question is, what restricted it in the first place? The cessation of Indian management, establishment of European homesteading and fire suppression in the late 1800s were favoured as the causes of the change, but it has been demonstrated that aspen was on the move south decades before the homesteaders arrived. Instead, it seems that the hand of Buffalo Bill set the aspen travelling: aspen pollen recovered from lakes in Canada indicate that the trees began invading the grassland in the 1870s, the very decade that bison reached their nadir. The bison, it seems, kept aspen in check by intensively browsing shoots and toppling mature trees by using them as scratching posts.[8]

Aspen is not the only tree affected by large mammals. Browsing by native herbivores can eliminate seed production in at least four species of willow. Balsam fir on Isle Royale suffered the same fate at the lips of moose. Dog rose and Canada yew are also hampered where native ungulates are numerous, and various berries and shrubs are under threat in national parks from overbrowsing. 'Ungulate browsing,' says one researcher, 'in recent times has virtually eliminated shrub seed production in areas frequented by wintering elk. Shrubs measured inside exclosures produced up to 20,000 times more berries than unprotected plants. These circumstances are not unique to Yellowstone and similar situations exist in other western states.'[9]

Two inadequate remedies are often proposed to avert this growing biological disaster: to manage the land in order to maintain it as it was when the Indians lived on it, or to leave it alone—to treat it as a wilderness—in the hope that it will return to some sort of ecological balance. How one might mimic Indian management is unclear and is itself a topic of debate, but to do nothing is also an impractical and perilous venture.

If such issues are ever to be addressed meaningfully, the future of North America's reserved lands needs to become a broad and magnificent debate that attempts to deal with the heart of the problem: ever since the extinction of the megafauna 13,000 years ago, the continent has had a seriously unbalanced fauna. The problem with elk may be a lack of the right predators. Wolves have already been reintroduced into Yellowstone and cougar remain, but do ambush predators such as jaguar and lion need to be reintroduced as well? These native species would certainly protect berry-producing bushes as well as aspen, for any areas that provide cover for large felines would be avoided by elk. The interminable debates about how many elk and wolf there should be are thus simply not enough.

Aldo Leopold died while fighting a brush fire in Wisconsin in 1948. It was an ironic death, for wildfires often result from the great problem that Leopold identified and devoted his life to righting—that of the ecological integrity of the United States. Fire behaves much like mega-herbivores do—consuming dry and coarse vegetation that smaller creatures, being more dainty eaters, turn their nose up at. Competition between large creatures, such as elephants and bison, and fire does exist, with fire being less frequent and intense where such herbivores live in large numbers. Take the cows out of a paddock, for example, and the grass grows until it yellows and dries. Given the right conditions, fire will then replace the cows as the recycler of the vegetation. And so fire management is as much part of the debate surrounding the vanished megafauna as is the demise of aspen.

I believe that the great question faced by park managers in North America today is whether, where it is suitable, they should reintroduce elephant, camel, Chacoan peccary, llama, panther and lion into their reserves. It is important to consider the truly gigantic species, for as G. Evelyn Hutchinson said in his 'Hints for an Agenda', if one has to prioritise, it's good to know 'how big is it and how fast does it happen?' As far as megafauna is concerned, we could begin with considering the two living species of elephant and their suitability as ecological replacements for the mammoth (the Indian elephant is closest) and the mastodon (the African elephant bears some similarities).[10]

One key question is whether megafauna can work for humanity. If so, then perhaps the animals have a role to play in the managed rangelands as well as in reserves. In much of the continent we have gone as far as we can with agricultural mass production using its European components in a traditional manner. Unless there is dramatic change soon agriculture must decline in large areas of the west, while elsewhere the relationship between North Americans and their land must also change if sustainability is to be achieved. Just how the most fragile lands might be managed sustainably is being tentatively addressed in America's heartland.

Having crossed the continent, the first wave of the land frontier rebounded back east, and in the late nineteenth century washed up on the Great Plains. So inhospitable is this region that homesteaders referred to it as the 'Great American Desert'. Still, when all other land was taken, land-hungry settlers took up even this difficult country and tried to make it pay. They were not the first agriculturalists to try to make a living there. About 1000 years ago village-dwelling Indians had moved out onto its vast expanse from the woodland margin. They advanced in a broad front, from the Dakotas to Texas, growing their cold-resistant varieties of maize and beans, retreating into stockades when threatened. They too suffered massacres and scalping, probably from hostile hunter-gatherer tribes similar to those fought by the US cavalry. Drought seems to have forced many of these Indian sod-busting pioneers to abandon their farms in the fifteenth century. Just over a century later the arrival of the horse tipped the balance in favour of a buffalo-hunting existence, leaving the fields fallow until the coming of the European sod-busters.[11]

Today, across some 362,000 square kilometres of the Great Plains, the European inhabitants of this tough country are facing the harsh reality of the region. The land is too poor to provide a financial base to support the complexity and richness of contemporary North American culture. The Europeans survive at lower population densities than did the Indians a century and a half ago. It has been seriously proposed that this enormous region should be left to the buffalo, which may provide subsistence for a sort of hunter-gathering ranching enterprise.[12]

Perhaps a century from now elephants will once again roam North America, together with large numbers of bison, llama, tapir, jaguar, camel and Chacoan peccary. These could well form the nucleus of a smaller yet sustainable economy, providing ecologically inexpensive meat and hides to a new and finally adapting people. Such a pipedream may not be all that far away, for it has already been demonstrated that in dry rangelands such as the Sonoran Desert a diverse assemblage of introduced browsers and grazers can produce higher yields per hectare to ranchers than can cattle alone.[13]

In contemplating these momentous changes to North America's ecosystems there is no need to fear outrageous ecological disruption, for we must remember that with the exception of humans new large mammal immigrants into North America have rarely if ever caused extinctions in the existing fauna. Also, the species discussed here are all close relatives of, if not identical with, America's original megafauna, which was exiled from the continent a mere 13 millennia ago, an eye blink in the grandeur of time. It is too short a period for any new mammal species to have evolved, or for ecosystems to reach a new equilibrium that excludes megafauna. We must also remember that mammal immigration into North America has been a great constant over the last 65 million years.

Curiously, at least one ice-age giant may return to North America even without the help of humans. Fourteen thousand years ago much of North America south of the latitude of New York was inhabited by a giant armadillo known as *Dasypus bellus*. At about twenty kilograms it was three times as heavy as the living armadillo, but otherwise the two were virtually identical. Beginning about 13,000 years ago the giant armadillo was driven out of what is now the United States. Human hunting was likely responsible, for the large armadillos probably hibernated in fissures among rocks, where they were easily spotted and dispatched. Some armadillos, however, survived in the dense rainforests and rugged terrain of Central America. These were smaller than those living further north, and they, like many species, have shrunk in average body size over the past 13,000 years. This is because they were hunted assiduously by the Indians who considered them a delicious repast and, as we have seen, hunting pressure can lead to the selection of rapidly maturing dwarfs.

As the Indian population itself began to decline in the seventeenth century and hunting pressure eased, the selective pressure felt by the armadillos reversed. Small armadillos began a rapid advance north. Between 1905 and 1914, at the height of their northward charge, they advanced at a rate of more than twenty-five kilometres per year. Today's small armadillos have still not reached as far north as their giant ancestors did 13,000 years ago. This is because they are more likely to die during cold winters than larger armadillos—they cannot accumulate as much fat and they lose heat more rapidly. As cold winters kill the smaller individuals, however, natural selection is acting upon the advance guard in the armadillo invasion and they are again increasing in size.[14] If the trend continues, in a few centuries from now the giant *Dasypus bellus* may once again stalk Appalachian forests and thrive as far north as New York.

The time has come now to turn our back on the past, to summarise what we have learned of the forces shaping this New World and to look towards the great issues and dilemmas that the North Americans will have to face in coming centuries.

TWENTY-SEVEN

REINVENTING AMERICA

We have seen that, upon entering a new homeland, all immigrant species are affected by three evolutionary forces—the founder effect, ecological (and social) release, and adaptation. We can think of these three forces as forming a great trajectory, like an upward arrow shot from a bow. The founder effect is the point of origin, ecological release the rapid rise of the arrow until it reaches its zenith, and adaptation its slow fall back to Mother Earth.

These forces have been acting on all immigrant species from the earliest times, but are difficult to discern in the fossil record. We can see their effects in human societies because so much is recorded, yet here the impact of evolutionary forces is masked by the unparalleled power of human societies to alter their own destiny. Furthermore, these three forces pull societies in different directions at different times, even reversing the tide of change as the strength of one force is overtaken by another. In most literate human societies the most powerful of these forces is arguably the founder effect. This is because words—especially written words—bind us in a way that nothing else can.

The importance of the founder effect in North American society might be deduced from the fact that its European-derived population has grown rapidly from a very small base. Any biological founder effect, however, was swamped long ago by immigration. Yet this immigration has not diluted the social founder effect, for each wave of immigrants has itself been shaped by a set of social compacts that determines relationships between industry, politics and the law, which in turn determines the distribution of wealth in society.

In the United States the Constitution has been the principal force behind the social founder effect. Despite twenty-seven amendments (eleven of which occurred in the eighteenth century; and one of which served to repeal another), the Constitution has proved remarkably resistant to change. It has guided political life for over two centuries and this founding compact has been the basis for the political, economic and spiritual structures into which all migrants are fitted. Thus while US society has grown far beyond the bounds of those who set the conventions, the founding compacts have been difficult to alter. This is in part because built-in protective mechanisms make change difficult and because the diffuse balance of power that exists in contemporary societies makes it hard to reshape.

The second force, one of release (both ecological and social) is, in translocated animal and human societies with abundant resources, extremely powerful. Indeed, despite the power of the founder effect, through the medium of the historic frontier the release force has been the most powerful in shaping North American society. In the seventeenth and eighteenth centuries Europeans migrating to North America were often escaping social, political and religious tyranny. In effect their arrival in the New World marked a moment of both social and ecological release, which was amplified on the frontier. The hatred of tyranny, so often expressed in early American writings, along with an egalitarian nature, are the most palpable effects of this release. The affluent, egalitarian society that North Americans had created by the nineteenth century thus resulted from social and ecological release. It was a liberty that was clearly in conflict with the concept of slavery, and this great contradiction was to find its own bloody resolution.

The third force operating on immigrant societies is adaptation to their new homeland. For humans, adaptation often comes about through modifying their cultural practices in order to accommodate the constraints of their new environment. In essence, adaptation is the opposite of release. Societal changes resulting from release are *not* adaptations to one's new homeland, for by definition they result only from escaping the constraints of the old, while adaptation is a response to the constraints of the new. Because North America is such a rich continent, Europeans have as yet experienced very little adaptation caused by environmental constraint. Such adaptation typically occurs when the frontier terminates. Despite its limited effects thus far, it is an important force because it produces the truly local forms of life that characterise societies and species. It is also the only force capable of producing, in an ecological sense, a diverse and adapted North American society.

The 'American society' identified by writers such as Frederick Jackson Turner was created largely from ecological and social release on the frontier. Release can produce a non-European society, but it cannot produce a society genuinely adapted to North American conditions. That can only exist after the frontier closes.

'He would be a rash prophet who would assert that the expansive character of American life has now entirely ceased,' wrote Turner in 1893. The idea that America and expansion are synonymous has deep roots. Even Benjamin Franklin believed that 'growth, expansion, and multiplication were the law of American life'. If such views are correct, North America as we know it may not have a long course to run, for all frontiers, no matter how extensive or productive, must end. At the dawn of the twenty-first century, with frontiers closing all around us, we must ask what will become of North America in what Marc Reisner has called this 'Era of Limits'.[1]

When Reisner was researching *Cadillac Desert* in the early 1990s he interviewed retired New Mexico state engineer Steve Reynolds. Reynolds had spent his life extending the land frontier by bringing water to the desert, creating hydroelectric power along the way. In the end he faced a lack of new rivers to dam as well as a crisis of public confidence. In his interview with Reisner, Reynolds turned to what in his mind was the solution to the problem—microwave satellites in space beaming power back to Earth. 'I've never felt that we should give up on space,' he said. 'It's our last frontier, and we need one.'[2]

The trouble, in part at least, is that all of us remain in love with the frontier, and with the image of ourselves as its free and heroic creatures; and yet unless a glimmer of that unrestrained era of growth can be chased through vending ever newer technologies to the world, the frontier is surely finished. What is worse, the limits that will characterise the centuries to come concern population as well as resources. Population growth is steadily slowing and global population looks set to stabilise or even reverse during the twenty-first century. What will this mean for the global economy? For better or worse, the 'American system' of manufacture and financing is now a global phenomenon, embracing even the most remote tribes. Yet the system has a profound weakness, for it was born on the frontier.

Few economists have addressed the issue, but as long ago as 1935 John Maynard Keynes looked into the future and did not like what he saw. He described his vision as the 'ultimate perplexity' of the system he had spent a lifetime studying. 'What will you do,' he asked, 'when you have built all the houses and roads and town halls and electric grids and water supplies and so forth which the stationary population of the future require?' At the heart of Keynes's dilemma lies the fact that in a world of shrinking and ageing population, capital itself can lose its value, for 'capital is not a self-subsistent entity existing apart from consumption'.[3] In other words, without increasing consumption, capital can have no increasing value.

Keynes's problem is one of limits, of terminating frontiers and of the degree to which economic growth is dependent upon population growth. The whole world has become part of the frontier that opened in North America in 1607, and we will all feel the effect of its closure. What shape the global economy will then take is unimaginable.

Leaving aside these greater questions, we can still ask how North America will change as it loses its frontiers. For the last 33 million years of Earth history Eurasia has been the world's sole 'ecological super-power'. Its creations have repeatedly invaded North America and the other lands within its reach, and have repeatedly reshaped North American ecosystems in their own ecological image. This situation has persisted throughout the human occupation of the continent, and it is only over the last fifty years—out of a span of 13,000—that North America has been the most influential continent on the planet. To a human being fifty years is everything—the best part of a lifetime—but in the scale of these events it is nothing, a mere grain in the Sahara of time.

The fifty years of US pre-eminence have come at a high price, for they have cost the continent much of its natural wealth and ecological stability. Even now aggressive capitalism is sacrificing the rivers, soils and poorer people of North America at the altar of the god of fortune, just as the Aztecs did their victims half a millennium ago.

Does an unalterable ethic—an economic, political and social culture of frontier over-exploitation—lie at the heart of North American society? Even now is some American, who has just entertained the possibility of reintroducing elephants to the prairie, dreaming of exporting giant elephant burgers to the world? I think not, as one of the most striking aspects of the North American people is their ability to reinvent themselves. The Puritans altered the ideals that were their very *raison d'être* as they made their first fortunes. Later those wishing to live as proud and free frontiersmen, who would not be shackled to the drudgery of the factory and who despised the burgeoning industrialisation, would invent the production line. Likewise the pioneers of the west worshipped rugged individualism, yet they soon embraced the socialism offered by federally funded water and agricultural schemes.

If the frontier dreaming of North America has to be destroyed so its environment and people can move into the future, then I'm sure it will be done. And there are signs that this is already occurring. After all, the frontier is a state of mind as much as anything, and even now the minds of its citizens are changing rapidly. Environmental protection is popular even with some of the conservative right, and is slowly closing what remains of the land, water, timber and fisheries frontiers of North America before complete disaster ensues.

One winter afternoon early in 1999 as I drove through central Pennsylvania, I pondered on how the titanic forces that have shaped the fates of the North American nations will exert themselves over the next millennium. A snowstorm had begun, and was predicted to turn to ice. All day I had been travelling the backroads of an idyllic countryside. The small farms of Mennonite families lay under a gentle dusting of snow, while handmade quilts, furniture and basketry were offered for sale at roadside markets. I had slowed to make way for the horse-drawn buggies, stealing glimpses at bearded men in straw boaters, their reins held taut as they hurried their high-headed horses past roaring automobiles.

Now the snow came more thickly and I began to worry about driving on ice. I stopped for coffee at a small shop in the village of Intercourse. There, three Mennonite men stood, passing the time of day. They joined me as I chatted to the shopkeeper, and I asked them which was the best way south in such dismal weather. Where was I going? 'Georgia,' I said, and they looked at me silently. Finally one suggested that Route 30 might be best. It was only after I left them that I realised the insensitivity of my question. During a lifetime they would never have cause to drive their buggies as far as I proposed to travel in a few hours, and they certainly could not drive them along the freeways I asked after. All day, I realised, I had been unconsciously regarding the Mennonites as actors such as one might encounter in a historical theme park. For them, horses and buggies, quilts and quaint clothing are not part of a performance but simply how they live. Less than three hours later I was approaching Washington, cocooned in a warm car. A signpost proclaimed 'NASA Center, Next Right'.

The Mennonites' isolationist response to their new homeland is hardly typical, for the majority of immigrants to the United States were transformed into citizens of that distinctive nation who seek not separation but inclusion within American popular culture. Because of the disparate origins of its peoples this popular culture must be stronger and able to stick more diverse parts together than is normally required from a cultural adhesive. In a sense American popular culture has to be a superglue, and to do that it must appeal to the lowest common denominator. American culture thus has no choice but to be mass culture, from baseball and football to hot dogs, Coca-Cola, McDonald's, baseball caps, pop tunes and Hollywood movies. It will always be superficial, full of self-reference to big-screen cliches, to soaps and their stars, and to a consensus news media. That's why it appeals to everyone, particularly teenagers, and that's why American popular culture is now a global phenomenon. Canadians, curiously, have never developed this kind of 'critical mass' culture that so thoroughly remakes its citizens. Instead a pluralism seems to lie at the heart of the Canadian ideal.

American political and cultural unity has served the nation well, but can we expect, as Abraham Lincoln first asked 140 years ago, that

American unity will hold long into the future? If the past is any guide, the British American culture born on the frontier will, like Clovis, diversify through time as people adapt to local conditions. In a hundred years Canada, and possibly even the United States, may resemble contemporary Europe—a plethora of distinctive provinces held together in a very loose federation.

Let us turn our gaze for a moment deep into the future and imagine, in the light of this ecological history, what the United States might be like in 1000 years. Given the sensitivity of the North American continent to climate change, the greenhouse effect is of special interest, for North America will feel its effects more violently and well in advance of other continents. Many climatologists anticipate rapid warming over the next century. When the Earth warmed by around one degree Celsius some 4000 years ago, the Indian societies inhabiting the south-west collapsed, leaving a parched, hot and mostly uninhabitable landscape. As yet little is known of changes in seasonal extremes or cyclone intensities during this time, but such things can be expected to have altered as well.

A significant minority of climatologists think that global warming will, paradoxically, hail another advance of the ice. They argue that warming will increase snowfall at the poles, and that if enough snow falls it will not melt completely over summer. As the snow accumulates it will reflect part of the sun's light back into space, causing rapid cooling to follow on the heels of the warming trend, with more snow accumulation and the growth of glaciers. Such an outcome would be catastrophic for North America. Canada would cease to exist as a human habitat, its people forced to migrate or invade another country. The United States would also suffer, for its northern cities such as Boston and New York could be bulldozed by glaciers and ground to loess. The deserts of the south-west, however, would bloom as the north and east froze.

While a future so distant may seem unknowable or unimaginable, other more ordinary outcomes can be guessed at. I think I know, for example, what most Americans will look like 1000 years from now. The majority will, I believe, resemble the Americans who occupied the continent before the historic frontier ever opened in 1492. Why do I think this? Because for the last few thousand years Mexico, with its dense population, has been the great powerhouse of cultural diffusion in North America, and it is still a mass exporter of people. Today's Mexicans are the fruit of the ecological bargain struck between the conquistadors and the inhabitants of the Aztec empire. As such they retain more native American culture and genes than any other group on the continent. If current trends continue it is these people who will achieve numerical dominance, at least in the United States, in the near future. The New World will then have come full circle.

If this comes to pass, then for North Americans of 1000 years hence the frontier will be a distant mirage, its influence long attenuated or vanished. What will their society be like? The amazing Frederick Jackson Turner knew, at least in outline, for in 1893 he wrote that when the frontier closes, 'the stubborn American environment [will be] there with its imperious summons to accept its conditions'.[4]

Just what those conditions are, and how people respond to them, awaits discovery by these almond-eyed children of Malinche. I wish them well in their struggle to survive in their remarkable land.

NOTES

INTRODUCTION

1 Porter, G., foreword, Hounshell, D. A., *From the American System to Mass Production, 1800–1932*, Johns Hopkins University Press, Baltimore, 1984.

2 Obradovich, J. D., 'A Cretaceous Time Scale' in *Evolution of the Western Interior Basin*, Geological Association of Canada special paper no. 39, W. G. E. Caldwell & E. G. Kauffman eds, 1993, pp. 379–96.

ACT 1: 66–59 MILLION YEARS AGO, IN WHICH AMERICA IS CREATED AND UNDONE

ONE: GROUND ZERO

1 Prothero, D. R., 'The Chronological, Climatic and Paleogeographic Background to North American Mammal Evolution', chapter 1 in *Evolution of Tertiary Mammals of North America, Volume 1: Terrestrial Carnivores, Ungulates, and Ungulate-Like Mammals*, C. M. Janis, K. M. Scott & L. L. Jacobs eds, Cambridge University Press, Cambridge, 1998.

2 Behler, J. L. & King, F. W., *National Audubon Society Field Guide to North American Reptiles and Amphibians*, 16th edn, Alfred A. Knopf, New York, 1998.

3 Wynn, J. C. & Shoemaker, E. M., 'The Day the Sands Caught Fire' in *Scientific American*, Nov. 1998, pp. 65–71.

4 Kyte, F. T., 'A Meteorite from the Cretaceous-Tertiary Boundary' in *Nature* vol. 396, 1998, pp. 237–39.

5 ibid.
   Shukolyukov, A. & Lugmair, G. W., 'Isotopic Evidence for the Cretaceous-Tertiary Impactor and Its Type' in *Science* vol. 282, 1998, pp. 927–29.

6 Melosh, H. J., Schneider, N. M., Zahnle, K. J. & Latham, D., 'Ignition of Global Wildfires at the Cretaceous-Tertiary Boundary' in *Nature* vol. 343, 1990, pp. 251–54.
   Schultz, P. H. & D'Hondt, S. D., 'Cretaceous-Tertiary (Chicxulub) Impact Angle and Its Consequences' in *Geology* vol. 24, 1996, pp. 963–67.

7 Landis, G. P., Rigby, J. K., Sloan, R. E., Hengst, R. & Snee, L. W., 'Pele Hypothesis: Ancient Atmospheres and Geologic-Geochemical Controls on Evolution, Survival, and Extinction' in *Cretaceous-Tertiary Mass Extinctions: Biotic and Environmental Changes*, N. Macleod & G. Keller eds, W. W. Norton & Co., New York, 1996, pp. 519–56.

8 Prothero, D. R., 'Protoceratidae', chapter 29 in *Evolution of Tertiary Mammals of North America*, op. cit.

Smit, J., 'Extinction and Evolution of Planktonic Foraminifera at the Cretaceous-Tertiary Boundary after a Major Impact' in *Geological Society of America special paper no. 190*, 1994, pp. 329–52.

9 Ward, P. D., *The Call of Distant Mammoths*, Copernicus Springer-Verlag, New York, 1997.

10 Thornton, I., *Krakatau: The Destruction and Reassembly of an Island Ecosystem*, Harvard University Press, Cambridge, 1996.

11 Johnson, K. R., 'Leaf-Fossil Evidence for Extensive Floral Extinction at the Cretaceous-Tertiary Boundary, North Dakota, USA' in *Cretaceous Research* vol. 13, 1992, pp. 91–117.

Hanson, T. A., Farrell, B. R. & Banks Upshaw III, 'The First Two Million Years after the Cretaceous-Tertiary Boundary in East Texas: Rate and Paleoecology of the Molluscan Recovery' in *Paleobiology* vol. 19, 1993, pp. 251–65.

12 D'Hondt, S., Pilson, M. E. Q., Sigurdsson, H., Hanson Jr, A. K. & Carey, S., 'Surface-Water Acidification and Extinction at the Cretaceous-Tertiary Boundary' in *Geology* vol. 22, 1994, pp. 983–86.

13 Ward, P. D., *The Call of Distant Mammoths*, op. cit.

14 Sheehan, P. M., Fastovsky, D. E., Hoffmann, R. G., Berghaus, C. B. & Gabriel, D. L., 'Sudden Extinction of the Dinosaurs: Latest Cretaceous, Upper Great Plains, USA' in *Nature* vol. 254, 1991, pp. 835–38.

15 Askin, R. A. & Jacobson, S. R., 'Palynological Change across the Cretaceous-Tertiary Boundary on Seymour Island, Antarctica: Environmental and Depositional Factors' in *Cretaceous-Tertiary Mass Extinctions*, op. cit.

TWO: THE REORDERING OF NORTH AMERICA

1 Schultz, P. H. & D'Hondt, S. D., 'Cretaceous-Tertiary (Chicxulub) Impact Angle and Its Consequences' in *Geology* vol. 24, 1996, pp. 963–67.

2 Thornton, I., *Krakatau: The Destruction and Reassembly of an Island Ecosystem*, Harvard University Press, Cambridge, 1996.

3 Johnson, K. R., Nichols, D. J., Attrep, M. & Orth, C. J., 'High-Resolution Leaf-Fossil Record Spanning the Cretaceous-Tertiary Boundary' in *Nature* vol. 340, 1989, pp. 708–11.

4 Wolfe, J. A. & Upchurch, G. R., 'Leaf Assemblages across the Cretaceous-Tertiary Boundary in the Raton Basin, New Mexico and Colorado' in *Proceedings National Academy Science* vol. 84, 1987, pp. 5096–5100.

5 Brouwers, E. M. & De Decker, P., 'Earliest Origin of Northern Hemisphere Temperate Nonmarine Ostracode Taxa: Evolutionary Development and Survival through the Cretaceous-Tertiary Boundary Mass-Extinction Event' in *Cretaceous-Tertiary*

Mass Extinctions: Biotic and Environmental Changes, N. MacLeod & G. Keller eds, W. W. Norton & Co., New York, 1996.

6 Wolfe, J. A., 'North American Non-Marine Climates and Vegetation during the Late Cretaceous' in *Palaeogeography, Palaeoclimatology, Palaeoecology* vol. 61, 1987, pp. 33–77.

7 Wing, S. L., 'Tertiary Vegetation of North America As a Context for Mammalian Evolution', chapter 2 in *Evolution of Tertiary Mammals of North America, Volume 1: Terrestrial Carnivores, Ungulates, and Ungulate-Like Mammals*, C. M. Janis, K. M. Scott & L. L. Jacobs eds, Cambridge University Press, Cambridge, 1998.

8 Li, H-L., 'Metasequoia: A Living Fossil' in *American Scientist* vol. 52, 1964, pp. 93–109.

9 Donnelly, T. W., 'Mesozoic and Cenozoic Plate Evolution of the Caribbean Region', chapter 4 in *The Great American Biotic Interchange*, F. G. Stehli & S. D. Webb eds, Plenum Press, New York, 1985, pp. 89–122.

10 Wing, S. L., op. cit.

THREE: OF HELLBENDERS AND HOOFED CREATURES

1 Lillegraven, J. A., 'Evolution of Wyoming's Paleocene Eastern Lowlands Related to Tectonic Loading by the Cannonball Sea', Society of Vertebrate Paleontology Abstracts, 1997, p. 60a.
    Prothero, D. R., 'The Chronological, Climatic and Paleogeographic Background to North American Mammal Evolution', chapter 1 in *Evolution of Tertiary Mammals of North America, Volume 1: Terrestrial Carnivores, Ungulates, and Ungulate-Like Mammals*, C. M. Janis, K. M. Scott, & L. L. Jacobs eds, Cambridge University Press, Cambridge, 1998.

2 & 3 Bryant, L. J., 'Non Dinosaurian Lower Vertebrates across the Cretaceous-Tertiary Boundary in Northeastern Montana' in *University of California Publications in Geological Sciences* vol. 134, 1989.

4 Estes, R. & Báez, A., 'Herpetofaunas of North and South America during the Late Cretaceous and Cenozoic: Evidence for Interchange?', chapter 6 in *The Great American Biotic Interchange*, F. G. Stehli & S. D. Webb eds, Plenum Press, New York, 1985, pp. 140–200.

5 & 6 Bryant, L. J., op. cit.

7 & 8 Behler, J. L. & King, F. W., *National Audubon Society Field Guide to North American Reptiles and Amphibians*, 16th edn, Alfred A. Knopf, New York, 1998.

9 Hartman, J. H., 'The Biostratigraphy and Palaeontology of Latest Cretaceous and Paleocene Freshwater Bivalves from the Western Williston Basin, Montana, USA' in

Bivalves: An Eon of Evolution, P. A. Johnston & J. W. Haggart eds, University of Calgary Press, Calgary, 1988.

10 Lofgren, D. L., 'The Bug Creek Problem and the Cretaceous-Tertiary Transition at McGuire Creek, Montana' in *University of California Publications in Geological Sciences* vol. 40, 1995, pp. 1–184.

11 Archibald, J. D., 'A Study of Mammalia and Geology across the Cretaceous-Tertiary Boundary in Garfield County, Montana' in *University of California Publications in Geological Sciences* vol. 122, 1982.

12 Ting, S., 'Paleocene and Early Eocene Land Mammal Ages of Asia' in *Dawn of the Age of Mammals in Asia*, K. C. Beard & M. R. Dawson eds, *Bulletin of the Carnegie Museum of Natural History* no. 34, Pittsburgh, 1998, pp. 124–47.

13 ibid.
   Lofgren, D. L., op. cit.
   Archibald, J. D., op. cit.

14 ibid.

15 Lillegraven, J. A. & Eberle, J. J., 'Vertebrate Faunal Changes through Lancian and Puercan Time in Southern Wyoming' in *Journal of Paleontology* vol. 73 (4), 1999, pp. 691–710.

16 Benson, R. D., '*Presbyornis isoni* and Other Late Paleocene Birds from North Dakota' in *Smithsonian Contribution to Palaeobiology* vol. 89, 2000, pp. 253–59.
   Hope, S., 'A New Species of *Graculavus* from the Cretaceous of Wyoming' in *Smithsonian Contribution to Palaeobiology* vol. 89, 2000, pp. 261–65.

17 Lillegraven, J. A. & Eberle, J. J., op. cit.

18 Maas, M. C. & Krause, D. W., 'Mammalian Turnover and Community Structure in the Paleocene of North America' in *Historical Biology* vol. 8, 1994, pp. 91–128.

19 Gingerich, P. D., 'South American Mammals in the Paleocene of North America', chapter 5 in *The Great American Biotic Interchange*, op. cit, pp. 123–37.

20 Lucas, S. G., Schloch, R. M. & Williamson T. E., 'Taeniodonta', chapter 16 in *Evolution of Tertiary Mammals of North America*, op. cit.

ACT 2: 57–33 MILLION YEARS AGO, IN WHICH AMERICA
BECOMES A TROPICAL PARADISE

FOUR: FIRST CONTACTS

1 Lucas, S. G., 'Pantodonta', chapter 18 in *Evolution of Tertiary Mammals of North America, Volume 1: Terrestrial Carnivores, Ungulates, and Ungulate-Like Mammals*,

C. M. Janis, K. M. Scott, & L. L. Jacobs eds, Cambridge University Press, Cambridge, 1998.

2 Cifelli, R. L. & Schaff, C. R., 'Arctostylopida', chapter 21 in *Evolution of Tertiary Mammals of North America*, op. cit.

3 & 4 Gingerich, P. D., 'South American Mammals in the Paleocene of North America', chapter 5 in *The Great American Biotic Interchange*, F. G. Stehli & S. D. Webb eds, Plenum Press, New York, 1985, pp. 123–37.

5 Cifelli, R. L., 'The Origin and Affinities of the South American Condylarthra and Early Tertiary Litopterna (Mammalia)', *American Natural History Museum Novitates 2772*, 1983.

6 Beard, K. C., 'East of Eden: Asia As an Important Center of Taxonomic Origination in Mammalian Evolution' in *Dawn of the Age of Mammals in Asia*, K. C. Beard & M. R. Dawson eds, *Bulletin of the Carnegie Museum of Natural History* vol. 34, Pittsburgh, 1998, pp. 5–39.

7 Leidy, J., 'On Some New Species of Fossil Mammalia from Wyoming' in Proceedings of the Academy of Natural Sciences of Philadelphia, 1872, pp. 167–169.

8 & 9 Colbert, E. H., *The Great Dinosaur Hunters and Their Discoveries*, Dover, New York, 1984.

10 Prothero, D. R., *The Eocene-Oligocene Transition*, Columbia University Press, New York, 1994.

11 Colbert, E. H., *The Great Dinosaur Hunters and Their Discoveries*, op. cit.

12 Prothero, D. R., *The Eocene-Oligocene Transition*, op. cit.
    Marsh, O. C., 'On the Gigantic Fossil Mammals of the Order Dinocerata' in *American Journal of Science* vol. 3, 1873, pp. 117–122.
    Marsh, O. C., 'Dinocerata: A Monograph of an Extinct Order of Gigantic Mammals' in *Monograph of the United States Geological Survey* vol. 10, 1886.

13–16 Betts, C., 'The Yale College Expedition' in *Harper's New Monthly Magazine*, Oct. 1871, pp. 663–71.

17 Bourgeois, J., Hansen, T. A., Wiberg, P. L. & Kaufman, E. G., 'A Tsunami Deposit at the Cretaceous-Tertiary Boundary in Texas' in *Science* vol. 241, 1988, pp. 567–70.
    Webb, S. D., 'Main Pathways of Mammalian Diversification in North America', chapter 7 in *The Great American Biotic Interchange*, op. cit.
    Clyde, W. C. 'Effects of Latest Paleocene Climate Change and Immigration on Local Mammalian Communities in the Bighorn Basin, Wyoming: New Evidence from the McCulloch Peaks', Society of Vertebrate Paleontology Abstracts, 1997, p. 39a.

18 Censky, E. et al, 'Over-Water Dispersal of Lizards Due to Hurricanes' in *Nature* vol. 395, 1998, p. 556.

19 Beard, K. C., op. cit.

20 Maas, M. C. & Krause, D. W., 'Mammalian Turnover and Community Structure in the Paleocene of North America' in *Historical Biology* vol. 8, 1994, pp. 91–128.
    Buchardt, B., 'Oxygen Isotope Paleotemperatures from the Tertiary Period in the North Sea Area' in *Nature* vol. 275, 1979, pp. 121–23.
    Shackleton, N. J., 'Paleogene Stable Isotope Events' in *Palaeogeography, Palaeoclimatology, Palaeoecology* vol. 57, 1986, pp. 91–102.

21 Janis, C. M., 'Tertiary Mammal Evolution in the Context of Changing Climates, Vegetation, and Tectonic Events' in *Annual Review of Ecological Systematics* vol. 24, 1993, pp. 467–500.

FIVE: THE BRIDGE OVER GREENLAND

1 Webb, S. D., 'Main Pathways of Mammalian Diversification in North America', chapter 7 in *The Great American Biotic Interchange*, F. G. Stehli & S. D. Webb eds, Plenum Press, New York, 1985, pp. 201–18.

2 Witmer, L. W. & Rose, K. D., 'Biomechanics of the Jaw Apparatus of the Gigantic Eocene Bird *Diatryma*: Implications for Diet and Mode of Life' in *Paleobiology* vol. 17, 1991, pp. 95–120.

3 Webb, S. D., 'A History of Savanna Vertebrates in The New World' (part 1) in *Annual Review of Ecology and Systematics* vol. 8, 1977, pp. 355–81.

4 McKenna, M., 'Was Europe Connected Directly to North America Prior to the Mid-Eocene?' in *Evolutionary Biology* vol. 6, 1972, pp. 179–88.

5 Sibley, C. G. & Ahlquist, J. E., *Phylogeny and Classification of Birds*, Yale University Press, New Haven, 1990.

6 Schaal, S. & Ziegler, W. eds, *Messel: An Insight into the History of Life and of the Earth*, trans. by Monika Shaffer-Fehre, Clarendon Press, Oxford, 1992.

7 Webb, S. D., 'Main Pathways of Mammalian Diversification in North America', op. cit.

8 Thewissen, J. G. M., 'Evolution of Paleocene and Eocene Phenacodontidae (Mammalia: Condylarthra)' in *Papers in Paleontology* vol. 29, Museum of Paleontology, 1990, pp.1–107.
    Rose, K. D., 'On the Origin of the Order Artiodactyla' in Proceedings of the National Academy of Science USA vol. 93, 1996, pp. 1705–9.

9 Beard, K. C., 'East of Eden: Asia As an Important Center of Taxonomic Origination in Mammalian Evolution' in *Dawn of the Age of Mammals in Asia*, K. C. Beard & M. R. Dawson eds, *Bulletin of the Carnegie Museum of Natural History* vol. 34, Pittsburgh, 1998, pp. 5–39.

ACT 3: 32 MILLION TO 13,000 YEARS AGO, IN WHICH AMERICA BECOMES A LAND OF IMMIGRANTS

SIX: A FATAL CONFIGURATION

1 Diamond, J., *Guns, Germs and Steel: A Short History of Everybody for the Last 13,000 Years*, Jonathan Cape, London, 1997.

2 & 3 Reisner, M., *Cadillac Desert: The American West and Its Disappearing Water*, Viking, New York, 1986.

4 Black, C. C., 'Holarctic Evolution and Dispersal of Squirrels (Rodentia: Sciuridae)' in *Evolutionary Biology* vol. 6, T. Dobshansky, M. K. Hecht & W. C. Steere eds, Appleton-Century-Crofts, New York, 1972, pp. 305–21.

5 Burnham, R. T., Wing, S. L. & Parker, G. G., 'The Reflection of Deciduous Forest Communities in Leaf Litter: Implications for Autochthonous Litter Assemblages from the Fossil Record' in *Paleobiology* vol. 18, 1992, pp. 34–53.

6 Wolfe, J. A., 'Late Cretaceous-Cenozoic History of Deciduousness and the Terminal Cretaceous Event' in *Paleobiology* vol. 13, 1987, pp. 215–26.

SEVEN: LA GRANDE COUPURE

1 Prothero, D. R., *The Eocene-Oligocene Transition*, Columbia University Press, New York, 1994.

Miller, K. G., 'Middle Eocene to Oligocene Stable Isotopes, Climate and Deep Water History: The Terminal Eocene Event?' in *Eocene–Oligocene Climatic and Biotic Evolution*, D. R. Prothero & W. A. Beggren eds, Princeton University Press, 1992, pp 160–77.

2 Wolfe, J. A., 'Tertiary Climatic Fluctuations and Methods of Analysis on Tertiary Floras' in *Palaeogeography, Palaeoclimatology, Palaeoecology* vol. 9, 1971, pp. 27–57.

3 Wolfe, J. A., 'Tertiary Floras and Paleoclimates of the Northern Hemisphere' in *Land Plants: Notes for a Short Course*, T. W. Broadhead ed., University of Tennessee Dept. Geological Sciences Studies in Geology 15, 1986, pp. 182–96.

4 Prothero, D. R., 'The Chronological, Climatic and Paleogeographic Background to North American Mammal Evolution', chapter 1 in *Evolution of Tertiary Mammals of North America, Volume 1: Terrestrial Carnivores, Ungulates, and Ungulate-Like Mammals*, C. M. Janis, K. M. Scott, & L. L. Jacobs eds, Cambridge University Press, Cambridge, 1998.

5 Wing, S. L., 'Eocene and Oligocene Floras and Vegetation of the Rocky Mountains', *Annals of the Missouri Botanical Garden* vol. 74, 1987, pp. 748–84.

Prothero, D. R., *The Eocene-Oligocene Transition*, op. cit.

6 Axelrod, D. I., 'Contributions to the Neogene Paleobotany of Central California', *University of California Publications in Geological Sciences* vol. 121, 1980.

7 Prothero, D. R., *The Eocene-Oligocene Transition*, op. cit.

Gunnell, G. F., 'Creodonta', chapter 5 in *Evolution of Tertiary Mammals in North America*, op. cit.

8 Sundell, K. S., 'Oreodonts: Large Burrowing Mammals of the Oligocene', Society of Vertebrate Paleontology Abstracts, 1997, p. 80a.

9 Foss, S. E., 'Behavioural Interpretation of the Entelodontidae (Mammalia: Artiodactyla)', Society of Vertebrate Paleontology Abstracts, 1998, p. 42a.

10 Wang, X. & Tedford, R. H., 'Canidae', chapter 21 in *The Terrestrial-Oligocene Transition in North America*, D. R. Prothero & R. J. Emry eds, Cambridge University Press, 1996.

11 Bryant, H. N., 'Nimravidae', chapter 22 in *The Terrestrial-Oligocene Transition in North America*, op. cit.

Rose, K. D., 'On the Origin of the Order Artiodactyla', *Proceedings of the National Academy of Science U.S.A.* vol. 93, 1996, pp. 1705–9.

12 MacFadden, B. J., *Fossil Horses: Systematics, Paleobiology and Evolution of the Family Equidae*, Cambridge University Press, Cambridge, 1992.

Janis, C. M. et al, *Evolution of Tertiary Mammals of North America*, op. cit.

13 Prothero, D. R., 'Camelidae', chapter 28 in *The Terrestrial-Oligocene Transition in North America*, op. cit.

14 Callahan, J. E., 'Velocity Structure and Flux of the Antarctic Circumpolar Current' in *Journal of Geophysical Research* vol. 76, 1971, pp. 5859–70.

Prothero, D. R., *The Eocene-Oligocene Transition*, op. cit.

Webb, S. D., 'Main Pathways of Mammalian Diversification in North America', chapter 7 in *The Great American Biotic Interchange*, F. G. Stehli & S. D. Webb eds, Plenum Press, New York, 1985, pp. 201–18.

15 Wolfe, J. A., 'Tertiary Climatic Fluctuations and Methods of Analysis on Tertiary Floras' in *Palaeogeography, Palaeoclimatology, Palaeoecology* vol. 9, 1971, pp. 27–57.

Wolfe, J. A., 'A Paleobotanical Interpretation of Tertiary Climates in the Northern Hemisphere' in *American Scientist* vol. 66, 1978, pp. 694–703.

Wolfe, J. A., 'Distributions of Major Vegetation Types during the Tertiary' in *The Carbon Cycle and Atmospheric CO2: Natural Variations, Archean to Present*, E. T. Sundquist & W. S. Broeker eds, American Geophysical Union Geophysical Monographs vol. 32, 1985, pp. 357–76.

Wolfe, J. A., 'Climatic, Floristic and Vegetational Changes Near the Eocene-Oligocene Boundary in North America' in *Eocene-Oligocene Climatic and Biotic Evolution*, op. cit., pp. 412–436.

16 & 17 Prothero, D. R., *The Eocene-Oligocene Transition*, op. cit.

18 Sibley, C. G. & Ahlquist, J. E., *Phylogeny and Classification of Birds*, Yale University Press, New Haven, 1990.

19 Ehrlich, P. R., Dobkin, D. S. & Wheye, D., *The Birder's Handbook*, Simon & Schuster, New York, 1988.

20 Sibley, C. G. & Ahlquist, J. E., op. cit.

21 Holman, J. A., '*Texasophis galbraithi* New Species: The Earliest New World Colubrid Snake' in *Journal of Vertebrate Paleontology* vol. 3, 1983, pp. 223–25.

EIGHT: A GOLDEN AGE

1 Webb, S. D., 'The Rise and Fall of the Late Miocene Ungulate Fauna in North America' in *Coevolution*, M. H. Nitecki ed., University of Chicago Press, Chicago, 1983, pp. 267–307.

2 Janis, C. M., 'Tertiary Mammal Evolution in the Context of Changing Climates, Vegetation, and Tectonic Events' in *Annual Review of Ecological Systematics* vol. 24, 1993, pp. 467–500.

3 Janis, C. M., Scott, K. M. & Jacobs, L. L. eds, *Evolution of Tertiary Mammals of North America, Volume 1: Terrestrial Carnivores, Ungulates, and Ungulate-Like Mammals*, Cambridge University Press, Cambridge, 1998.

4 Owen-Smith, R. N., *Megaherbivores: The Influence of Very Large Body Size on Ecology*, Cambridge University Press, Cambridge, 1988.
    Voorhies, M. R., 'Ancient Ashfall Creates Pompeii of Ancient Animals' in *National Geographic*, Jan. 1981, pp. 66–75.
    Webb, S. D., 'The Rise and Fall of the Late Miocene Ungulate Fauna in North America', op. cit.

5 ibid.
    Janis, C. M. & Manning, E., 'Dromomerycidae', chapter 32 in *Evolution of Tertiary Mammals of North America*, op. cit.
    Webb, S. D., 'A History of Savanna Vertebrates in The New World' (part 1) in *Annual Review of Ecology and Systematics* vol. 8, 1977, pp. 355–81.

6 MacFadden, B. J., *Fossil Horses: Systematics, Paleobiology and Evolution of the Family Equidae*, Cambridge University Press, Cambridge, 1992.

7 Byers, J. A., *The American Pronghorn: Social Adaptations and the Ghosts of Predators Past*, University of Chicago Press, Chicago, 1997.

8 Webb, S. D., 'The Rise and Fall of the Late Miocene Ungulate Fauna in North America', op. cit.

9 Martin, L. D., 'Nimravidae', chapter 12 in *Evolution of Tertiary Mammals of North America*, op. cit.

10 Prothero, D. R., *The Eocene-Oligocene Transition*, Columbia University Press, New York, 1994.

11 Webb, S. D., 'Main Pathways of Mammalian Diversification in North America',

chapter 7 in *The Great American Biotic Interchange*, F. G. Stehli & S. D. Webb eds, Plenum Press, New York, 1985, pp. 201–18.

12 Hunt, R. M., 'Ursidae', chapter 10 in *Evolution of Tertiary Mammals of North America*, op. cit.

13 Prothero, D. R., 'The Chronological, Climatic and Paleogeographic Background to North American Mammal Evolution', chapter 1 in *Evolution of Tertiary Mammals of North America*, op. cit.

14 Barnosky, A. D., 'Age of the Mid-Tertiary Unconformity in the Western Rocky Mountains and Miocene Biogeography', Society of Vertebrate Paleontology Abstracts, 1998, p. 26a.

15 Prothero, D. R., 'The Chronological, Climatic and Paleogeographic Background to North American Mammal Evolution', op. cit.

16 Axelrod, D. I., 'Age and Origin of Sonoran Desert Vegetation' in *California Academy of Science Occasional Paper 132*, 1979, p. 72.
   Webb, S. D., 'A History of Savanna Vertebrates in The New World', op. cit.

NINE: GATEWAY TO THE PRESENT

1 Axelrod, D. I., 'Contributions to the Neogene Paleobotany of Central California' in *University of California Publications in Geological Sciences* vol. 121, 1980.

2 Prothero, D. R., 'Protoceratidae', chapter 29 in *Evolution of Tertiary Mammals of North America, Volume 1: Terrestrial Carnivores, Ungulates, and Ungulate-Like Mammals*, C. M. Janis, K. M. Scott, & L. L. Jacobs eds, Cambridge University Press, Cambridge, 1998.

3 Janis, C. M., 'Tertiary Mammal Evolution in the Context of Changing Climates, Vegetation, and Tectonic Events' in *Annual Review of Ecological Systematics* vol. 24, 1993, pp. 467–500.
   Webb, S. D., 'A History of Savanna Vertebrates in The New World' (part 1) in *Annual Review of Ecology and Systematics* vol. 8, 1977, pp. 355–81.

4 Axelrod, D. I., op. cit.

5 Prothero, D. R., 'The Chronological, Climatic and Paleogeographic Background to North American Mammal Evolution', chapter 1 in *Evolution of Tertiary Mammals of North America*, op. cit.

6 Kahlke, R. D., 'The History of the Origin, Evolution and Dispersal of the Late Pleistocene *Mammuthus Coelodonta* Faunal Complex in Eurasia (Large Mammals)', Mammoth Site Hot Springs, South Dakota, scientific papers, 1999.
   Voorhies, M. R. & Perkins, M. E., 'Odocoilene Deer (?*Bretzia*) from the Santee

Ash Locality, Late Hemphillian, Nebraska: Oldest New World Cervid?', Society of Vertebrate Paleontology Abstracts, 1998, p. 84a.

7 Flynn, L. J., Tedford, R. H. & Zhanxiang, Q., 'Enrichment and Stability in the Pliocene Mammalian Fauna of North China' in *Paleobiology* vol. 17, 1991, pp. 246–65.

Honey, J. G., Harrison, J. A., Prothero, D. R. & Stevens, M. S., 'Camelidae', chapter 30 in *Evolution of Tertiary Mammals of North America*, op. cit.

Tobien, H., 'Migrations of Proboscideans and Lagomorphs (Mammalia) across Bering Land Bridge in the Late Cenozoic' in *Beringia in the Cenozoic Era*, V. L. Kontrimavichus ed., Oxonian Press, New Delhi, 1984, pp. 327–38.

8 Flannery, T. F., *The Future Eaters: An Ecological History of the Australasian Lands and People*, Reed Books, Port Melbourne, 1994.

TEN: UNITED LANDS OF AMERICA

1 Iturralde-Vinent, M. A. & MacPhee, R. D. E., 'Paleogeography of the Caribbean Region: Implications for Cenozoic Biogeography' in *Bulletin of the American Museum of Natural History* vol. 238, 1999, pp. 1–95.

2 Gose, W. A., 'Caribbean Tectonics from a Paleomagnetic Perspective', chapter 11 in *The Great American Biotic Interchange*, F. G. Stehli & S. D. Webb eds, Plenum Press, New York, 1985, pp. 285–301.

Marton, G. & Buffler, R. T., 'Application of Simple Shear Model to the Evolution of Passive Continental Margins in the Gulf of Mexico Basin' in *Geology* vol. 21, 1993, pp. 495–98.

Smith, D. L., 'Caribbean Plate Relative Motions', chapter 2 in *The Great American Biotic Interchange*, op. cit., pp. 17–48.

3 Webb, S. D., 'Main Pathways of Mammalian Diversification in North America', chapter 7 in *The Great American Biotic Interchange*, op. cit., pp. 201–18.

4 & 5 Darlington, P. J., *Zoogeography: The Geographical Distribution of Animals*, John Wiley & Sons, New York, 1957.

6 Simpson, G. G., *Splendid Isolation: The Curious History of South American Mammals*, Yale University Press, New Haven, 1980.

7 Estes, R. & Báez, A., 'Herpetofaunas of North and South America during the Late Cretaceous and Cenozoic: Evidence for Interchange?', chapter 6 in *The Great American Biotic Interchange*, op. cit., pp. 140–200.

8 Marshall, L. G., 'Geochronology and Land-Mammal Biochronology of the Transamerican Faunal Interchange', chapter 3 in *The Great American Biotic Interchange*, op. cit., pp. 49–88.

9 Tambussi, C. P., 'Fororracoideos, las grandes aves carnívoras de la Patagonia de antaño' in *Museo*, 1999, pp. 61–65.

Brodkorb, P., 'A Giant Flightless Bird from the Pleistocene of Florida' in *The Auk* vol. 8, 1963, pp. 111–15.

10 Bussing, W. A., 'Patterns of Distribution of the Central American Ichthyofauna', chapter 17 in *The Great American Biotic Interchange*, op. cit., pp. 453–75.

Simpson, B. B. & Neff, J. L., 'Plants, Their Pollinating Bees, and the Great American Interchange', chapter 16 in *The Great American Biotic Interchange*, op. cit., pp. 427–52.

11 ibid.

Schultz, J. C. & Floyd, T., 'Desert Survivor' in *Natural History* vol. 108 (1), 1999, pp. 24–28.

12 Simpson, G. G., *Splendid Isolation*, op. cit.

Webb, S. D., 'Main Pathways of Mammalian Diversification in North America', op. cit.

ELEVEN: LAURENTIDE

1 Lurie, E., *Louis Agassiz: A Life in Science*, Johns Hopkins University Press, Baltimore, 1988.

2 Esat, T. M., McCulloch, M. T., Chappell, J., Pillans, B. & Omura, A., 'Rapid Fluctuations in Sea Level Recorded at Huon Peninsula during the Penultimate Glaciation' in *Science* vol. 283, 1999, pp. 197–201.

3 & 4 Flint, R. F., *Glacial and Quaternary Geology*, John Wiley & Sons, New York, 1971.

5 Esat, T. M. et al., op. cit.

6 COHMAP members, 'Climatic Changes of the Last 18,000 Years: Observations and Model Simulations' in *Science* vol. 241, 1988, pp. 1043–52.

7 Webb, T., 'The Appearance and Disappearance of Major Vegetational Assemblages: Long-Term Vegetational Dynamics in Eastern North America' in *Vegetatio* vol. 69, 1987, pp. 177–87.

8 Watts, W. A. & Hansen, B. C. S., 'Pre-Holocene and Holocene Pollen Records of Vegetation History from the Florida Peninsula and Their Climatic Implications', in *Palaeogeography, Palaeoclimatology, Palaeocology*, 1994.

9 Harris, A. H., *Late Pleistocene Vertebrate Paleoecology of the West*, University of Texas Press, Austin, 1985.

10 Martin, P. S., 'Vanishings, and Future, of the Prairie' in *Geoscience and Man* vol. 10, 1975, pp. 39–49.

11 Pyne, S. J., *Fire in America: A Cultural History of Wildland and Rural Fire*, University of Washington Press, Seattle, 1982.

TWELVE: VISIT TO A NEW WORLD

1 Owen-Smith, R. N., *Megaherbivores: The Influence of Very Large Body Size on*

Ecology, Cambridge University Press, Cambridge, 1988.

Shipman, P., 'Body Size and Broken Bones: Preliminary Interpretations of Proboscidean Remains', chapter 5 in *Proboscidea and Paleoindian Interactions*, J. W. Fox, C. B. Smith & K. T. Wilkins eds, Baylor University Press, Waco, 1992.

2 Kurten, B. & Anderson, E., *Pleistocene Mammals of North America*, Columbia University Press, New York, 1980.

3 Anderson, E., 'Who's Who in the Pleistocene: A Mammalian Bestiary' in *Quaternary Extinction: A Prehistoric Revolution*, P. S. Martin & R. G. Klein eds, University of Arizona Press, Arizona, 1984.

4 Graham, R. W., Farlow, J. O. & Vandike, J. E., 'Tracking Ice Age Fields: Identification of Tracks of *Panthera atrox* from a Cave in Southern Missouri, USA' in *Palaeoecology and Paleoenvironments of Late Cenozoic Mammals*, K. M. Stewart & K. L. Seymour eds, University of Toronto Press, Toronto, 1996, pp. 331–46.

5 Campbell, K. E. & Tonni, E. P., 'Preliminary Observations on the Paleobiology and Evolution of Teratorns (Aves: Teratornithidae)' in *Journal of Vertebrate Paleontology* vol. 1, 1981, pp. 265–72.

6 Emslie, S. D., 'The Fossil History and Phylogeneric Relationships of Condors (Ciconiformes: Vulturidae) in the New World' in *Journal of Vertebrate Paleontology* vol. 8, 1988, pp. 212–28.

7 Ehrlich, P. R., Dobkin, D. S. & Wheye, D., *The Birder's Handbook*, Simon & Schuster, New York, 1988.

8 Emslie, S. D. & Morgan, G. S., 'Taphonomy of a Late Pleistocene Carnivore Den, Dade County, Florida', chapter 6 in *Late Quaternary Environments and Deep History: A Tribute to Paul S. Martin*, D. W. Steadman & J. I. Mead eds, Mammoth Site Hot Springs, Cary, 1995.

9 Holman, J. A., *Pleistocene Amphibians and Reptiles in North America*, Oxford University Press, Cary, 1995.

10 & 11 Owen-Smith, R. N. & Danckwerts, J. E., 'Herbivory', chapter 17 in *The Vegetation of Southern Africa*, Cambridge University Press, Cambridge, 1997.

12 Guthrie, R. D., *Frozen Fauna of the Mammoth Steppe: The Story of Blue Babe*, University of Chicago Press, Chicago, 1990.

ACT 4: 11,000 YEARS AGO–AD 1491, IN WHICH AMERICA IS DISCOVERED

THIRTEEN: A NEW WORLD

1 Dillehay, T. D., *Monte Verde: A Late Pleistocene Settlement in Chile*, vol. 2, Smithsonian Institution Press, Washington, 1997.

2 Fox, J. W. & Smith, C. B., 'Introduction: Historical Background, Theoretical Approaches and Proboscideans' in *Proboscidean and Paleoindian Interaction*, J. W. Fox, C. B. Smith & K. T. Williams eds, Baylor University Press, Waco, 1992, pp. 1–15.

3 Guthrie, R. D., *Frozen Fauna of the Mammoth Steppe: The Story of Blue Babe*, University of Chicago Press, Chicago, 1990.

4 Fiedel, S. J., *Prehistory of the Americas*, Cambridge University Press, Cambridge, 2nd edn, 1993.

   Gamble, C., *The Palaeolithic Settlement of Europe*, Cambridge University Press, London, 1986.

5 Saunders, J. J., 'Blackwater Draw: Mammoths and Mammoth Hunters in the Terminal Pleistocene', chapter 8 in *Proboscidean and Paleoindian Interaction*, op. cit.

6 Fiedel, S. J., *Prehistory of the Americas*, op. cit.

7 Haynes, C. V., Beukins, R. P., Jull, A. J. T., & Davis, O. K., 'New Radiocarbon Dates for Some Old Folsom Sites: Accelerator Technology' in *Ice Age Hunters of the Rockies*, D. J. Stanford & J. S. Day eds, Denver Museum of Natural History and University Press, Colorado, 1992, pp. 83–100.

8 Frison, G., 'Clovis-Folsom-Goshen Relationships in the Northern High Plains' in *Megafauna and Man: Discovery of America's Heartland*, L. D. Agenboard, J. I. Mead & L. W. Nelson eds, Mammoth Site Hot Springs, South Dakota, scientific papers vol. 1, 1990, pp. 100–8.

9 Boorstin, D. J., *The Americans: The Colonial Experience*, Cardinal, London, 1958, reprinted 1991.

FOURTEEN: THE BLACK HOLE THEORY OF EXTINCION

1 Darwin, C., *Journal of Researches into the Geology and Natural History of the Various Countries Visited by H.M.S. Beagle*, (facsimile) Hafner, London, 1839.

   Wallace, A. R., *The Geographical Distribution of Animals: With a Study of the Relations of Living and Extinct Faunas As Elucidating Past Changes of the Earth's Surface*, Harper, New York, 1876.

2 Graham, R. W. & Lundelius, E. L., 'Coevolutionary Disequilibrium and Pleistocene Extinctions', chapter 11 in *Quaternary Extinctions: A Prehistoric Revolution*, P. S. Martin & R. G. Klein eds, University of Arizona Press, Tucson, 1984.

3 Grayson, D. K., 'The Chronology of North American Late Pleistocene Extinctions' in *Journal of Archaeological Science* vol. 16, 1989, pp. 153–65.

   Meltzer, D. J. & Mead, J. I., 'The Timing of Late Pleistocene Mammalian Extinctions in North America' in *Quaternary Research* vol. 19, 1983, pp. 130–35.

   Stafford, T., 'Late Pleistocene Megafaunal Extinctions and the Clovis Culture: Absolute Ages Based on Accelerator 14C Dating of Skeletal Remains' in *Megafauna and Man: Discovery of America's Heartland*, L. D. Agenboard, J. I. Mead &

L. W. Nelson eds, Mammoth Site of Hot Springs, vol. 1, 1990, pp. 118–22.

Steadman, D. W., Stafford, T. W. & Funk, R. E., 'Nonassociation of Paleoindians with AMS-Dated Late Pleistocene Mammals from the Dutchess Quarry Caves, New York' in Quaternary Research vol. 47, 1997, pp. 105–16.

4 Roberts, R. G., 'Luminescence Dating in Archaeology: From Origins to Optical', Radiation Measurements vol. 27, 1998, pp. 819–92.

5 Miller, G. H., Magee, J. W., Johnson, B. J., Fogel, M. L., Spooner, N. A., McCulloch, M. T. & Ayliffe, L. K., 'Pleistocene Extinction of Genyornis newtoni: Human Impact on Australian Megafauna' in Science vol. 283, 1999, pp. 205–8.

6 MacPhee, R. D. E., Flemming, C. & Lunde, D. P., '"Last Occurrence" of the Antillean Insectivoran Nesophontes: New Radiometric Dates and Their Interpretation', in American Museum Natural History Novitates 3261, 1999.

FIFTEEN: MASSACRING THE MAMMOTH,
DISMEMBERING THE MASTODON

1 Saunders, J. J., 'Blackwater Draw: Mammoths and Mammoth Hunters in the Terminal Pleistocene', chapter 8 in Proboscidean and Paleoindian Interaction, J. W. Fox, C. B. Smith & K. T. Williams eds, Baylor University Press, Waco, 1992.

2 Winter, W. H., 'Elephant Behaviour', East African Wildlife Journal no. 2, 1964, pp. 163–64.

3 Haynes, C. V., 'Elephant Hunting in North America' in Scientific American vol. 214 (6), 1966, pp. 104–12.

4 Martin, P. S. & Klein, R. G. eds, Quaternary Extinctions: A Prehistoric Revolution, University of Arizona Press, Tucson, 1984.

5 Owen-Smith, R. N., Megaherbivores: The Influence of Very Large Body Size on Ecology, Cambridge University Press, Cambridge, 1988.

Saunders, J. J., op. cit.

6 Fisher, D. C., 'Mastodon Procurement by Paleoindians of the Great Lakes Region: Hunting or Scavenging?' in The Evolution of Human Hunting, M. H. Nitecki & D. V. Nitecki eds, Plenum Press, New York, 1987, pp. 309–421.

Fisher, D. C. 'Experiments on Subaqueous Meat Caching', CPR, vol. 12, 1995, pp. 77–80.

Fisher, D. C., 'Testing Late Pleistocene Extinction Mechanisms with Data on Mastodon and Mammoth Life History', Society of Vertebrate Paleontology Abstracts vol. 16 (3), 1996, p. 34a.

Fisher, D. C., 'Extinctions of Proboscideans in North America', chapter 30 in The Proboscidea, J. Shoshani & P. Tassy eds, Museum of Paleontology, Michigan, 1997.

7 & 8 Laub, R. S., 'On Disassembling an Elephant: Anatomical Observations Bearing on Paleoindian Exploitation of Proboscidea', chapter 6 in *Proboscidean and Paleoindian Interaction*, op. cit.

9 Fisher, D. C., Society of Vertebrate Paleontology Abstracts, New York, October 1996.
   Ward, P. D., *The Call of Distant Mammoths*, Copernicus Springer-Verlag, New York, 1997.

10 Lepper, B. T., Frolking, T. A., Fisher, D. C., Goldstein, G., Sanger, J. E., Wymer, D. A., Ogden III, J. G. & Hooge, P. E., 'Intestinal Contents of a Late Pleistocene Mastodon from Midcontinental North America' in *Quaternary Research* vol. 36, 1991.

11 & 12 Fisher, D. C. 'Experiments on Subaqueous Meat Caching ', op. cit.

13 Owen-Smith, R. N., 'Pleistocene Extinctions: The Pivotal Role of Megaherbivores' in *Paleobiology* vol. 13, 1987, pp. 351–62.

SIXTEEN: THE NEW AMERICAN FAUNA

1 Guthrie, R. D., *Frozen Fauna of the Mammoth Steppe: The Story of Blue Babe*, University of Chicago Press, Chicago, 1990.

2 Elias, S. A., Short, S. K. & Birks, H. H., 'Late Wisconsin Environments on the Bering Land Bridge' in *Palaeogeography, Palaeoclimatology, Palaeoecology* vol. 136, 1997, pp. 293–308.

3 Oksanen, L., 'Ecosystem Organisation: Mutualism and Cybernetics or Plain Darwinian Struggle for Existence?' in *The American Naturalist* vol. 131, 1988, pp. 424–44.
   Owen-Smith, R. N., 'Pleistocene Extinctions: The Pivotal Role of Megaherbivores' in *Paleobiology* vol. 13, 1987, pp. 351–62.
   Wing, S. L., 'Tertiary Vegetation of North America As a Context for Mammalian Evolution', chapter 2 in *Evolution of Tertiary Mammals of North America, Volume 1: Terrestrial Carnivores, Ungulates, and Ungulate-Like Mammals*, C. M. Janis, K. M. Scott, & L. L. Jacobs eds, Cambridge University Press, Cambridge, 1998.

4 Graham, R. W. & Lundelius, E. L., 'Coevolutionary Disequilibrium and Pleistocene Extinctions', chapter 11 in *Quaternary Extinctions: A Prehistoric Revolution*, P. S. Martin & R. G. Klein eds, University of Arizona Press, Tucson, 1984.

5 Barnosky, A. D. et al., 'Comparison of Mammalian Response to Glacial-Interglacial Transitions in the Middle to Late Pleistocene' in *Palaeoecology and Palaeoenvironments of Late Cenozoic Mammals*, K. M. Stewart & K. L. Seymour eds, University of Toronto Press, Toronto, 1996.

6 Matthiessen, P., *Wildlife in America*, Penguin Nature Classics, Penguin Books, Harmondsworth, 1959.

7 Guthrie, R. D., 'Bison Evolution and Zoogeography in North America during the

Pleistocene' in *Quarterly Review of Biology* vol. 45, 1970, pp. 1–15.

8 Leonard, J. A., Wayne, R. K. & Cooper, A., 'Population Genetics of Ice Age Brown Bears', *PNAS 97*, 1999, pp. 1651–54.

9 Lister, A. M., 'Evolution of Mammoths and Moose: The Holarctic Perspective', chapter 9 in *Morphological Change in Quaternary Mammals of North America*, R. A. Martin & A. D. Barnosky eds, Cambridge University Press, New York, 1993.

10 Seymour, K., 'Size Change in North American Quaternary Jaguars', chapter 14 in *Morphological Change in Quaternary Mammals of North America*, op. cit.

11 Hampton, B., *The Great American Wolf*, Owl Books, New York, 1997.

12 & 13 Webb, S. D., (pers. comm.).

14 Graham, R., (pers. comm.).

15 Gamble, C., *The Palaeolithic Settlement of Europe*, Cambridge University Press, London, 1986.
      Kurten, B. & Anderson, E., *Pleistocene Mammals of North America*, Columbia University Press, New York, 1980.

SEVENTEEN: THE MAKING OF THE BUFFALO

1 & 2 Jones, R., 'Folsom and Talgai: Cowboy Archaeology in Two Continents', chapter 2 in *Approaching Australia: Papers from the Harvard Australian Studies Symposium*, H. Bolitho & C. Wallace-Crabbe eds, Harvard University Committee on Australian Studies, Cambridge, 1998.

3 Haynes, C. V., Beukins, R. P., Jull, A. J. T., & Davis, O. K., 'New Radiocarbon Dates for Some Old Folsom Sites: Accelerator Technology' in *Ice Age Hunters of the Rockies*, D. J. Stanford & J. S. Day eds, Denver Museum of Natural History & University Press, Colorado, 1992, pp. 83–100.

4 Hornaday, W. T., 'The Extermination of the American Bison', *Smithsonian Report*, Washington, 1889, pp. 369–548.

5 Fiedel, S. J., *Prehistory of the Americas*, Cambridge University Press, Cambridge, 2nd edn, 1993.

6 & 7 Guthrie, R. D., *Frozen Fauna of the Mammoth Steppe: The Story of Blue Babe*, University of Chicago Press, Chicago, 1990.

8 Frison, G. C., 'The Foothills-Mountains and the Open Plains: The Dichotomy in Paleoindian Subsistence Strategies between Two Ecosystems', chapter 9 in *Ice Age Hunters of the Rockies*, op. cit.

9 Pyne, S. J., *Fire in America: A Cultural History of Wildland and Rural Fire*, University of Washington Press, Seattle, 1982.

10 Gilbert, B. M. & Martin, L. D., 'Late Pleistocene Fossils of Natural Trap Cave, Wyoming, and the Climatic Model of Extinction', chapter 6 in *Quaternary Extinctions: A Prehistoric Revolution*, P. S. Martin & R. G. Klein

eds, University of Arizona Press, Tucson, 1984.

Wilson, M., 'Archaeological Kill Site Populations and the Holocene Evolution of the Genus *Bison*' in *Bison Evolution and Utilisation*, L. Davis & M. Wilson eds, *Plains Anthropologist* memoir no. 14, 1978, pp. 9–23.

11 Guthrie, R. D., 'Alaskan Megabucks, Megabulls and Megarams: The Issue of Pleistocene Gigantism', *Contributions in Quaternary Vertebrate Paleontology: A Volume in Memorial to John E. Guilday*, Carnegie Museum of Natural History, Pittsburgh, 1984.

12 Flannery, T. F., *The Future Eaters: An Ecological History of the Australasian Lands and People*, Reed Books, Port Melbourne, 1994.

EIGHTEEN: THE RISE OF CULTURES

1 Lewin, R., *Patterns in Evolution: The New Molecular View*, Scientific American Library, New York, 1999.

2 Wilson, J., *The Earth Shall Weep: A History of Native America*, Atlantic Monthly Press, New York, 1999.

3 Marsh, G. H. & Laughlin, W. S., 'Human Anatomical Knowledge among the Aleutian Islanders' in *Southwestern Journal of Anthropology* vol. 12, 1956, pp. 38–78.

4–8 Fiedel, S. J., *Prehistory of the Americas*, Cambridge University Press, Cambridge, 2nd edn, 1993.

9 Smith, J., *The True Travels, Adventures and Observations of Captain John Smith in Europe, Asia, Africa and America, and the General History of Virginia, New England and the Summer Isles, Books 1–3*, Cambridge University Press, Cambridge, republished 1908, pp. 1608–30.

10 Matthiessen, P., *Wildlife in America*, Penguin Nature Classics, Penguin Books, Harmondsworth, 1959.

11 Wilson, J., *The Earth Shall Weep*, op. cit.
Fiedel, S. J., *Prehistory of the Americas*, op. cit.

12 ibid.

13 & 14 Wilson, J., *The Earth Shall Weep*, op. cit.

15 Steadman, D. W. & Martin, P. S., 'Extinction of Birds in the Late Pleistocene of North America', chapter 21 in *Quaternary Extinctions: A Prehistoric Revolution*, P. S. Martin, & R. G. Klein, University of Arizona Press, Tucson, 1988.

16 Heizer, R. F. & Whipple, M. A., *The California Indians: A Source Book*, University of California Press, Berkeley, 1951.

17–20 Fiedel, S. J., *Prehistory of the Americas*, op. cit.

NINETEEN: THE TAMING OF TEOSINTE

1 Webb, S. D., 'Main Pathways of Mammalian Diversification in North America',

chapter 7 in *The Great American Biotic Interchange*, F. G. Stehli & S. D. Webb eds, Plenum Press, New York, 1985, pp. 201–18.

2 Fiedel, S. J., *Prehistory of the Americas*, Cambridge University Press, Cambridge, 2nd edn, 1993.

3 Pringle, H., 'The Slow Birth of Agriculture' in *Science* vol. 282, 20 Nov. 1998, pp. 1446–50.

4 Wilson, J., *The Earth Shall Weep: A History of Native America*, Atlantic Monthly Press, New York, 1999.

5 Pringle, H., op. cit.

6 & 7 Fiedel, S. J., *Prehistory of the Americas*, op. cit.

8 Diaz del Castillo, B., *The Discovery and Conquest of Mexico, 1517–1521*, trans. A. P. Maudslay, Grove Press, New York, 1958.

9 & 10 Fiedel, S. J., *Prehistory of the Americas*, op. cit.

ACT 5: 1492–2000, IN WHICH AMERICA CONQUERS THE WORLD

TWENTY: ALTERNATIVE AMERICAS

1 Kurlansky, M., *Cod: A Biography of the Fish That Changed the World*, Jonathan Cape, London, 1997.

2 Sauer, C. O., *Sixteenth Century North America*, University of California Press, Berkeley, 1975.
  Shenkman, R., *Legends, Lies and Cherished Myths of American History*, Harper-Perennial, New York, 1989.

3 Carleton, M. D. & Olson, S. L., 'Amerigo Vespucci and the Rat of Fernando de Noronha: A New Genus and Species of Rodentia (Muridae: Sigmodontinae) from a Volcanic Island off Brazil's Continental Shelf', *American Museum Natural History Novitates 3256*, 1999.

4 Diaz del Castillo, B., *The Discovery and Conquest of Mexico, 1517–1521*, trans. A. P. Maudslay, Grove Press, New York, 1958.

5 Farb, P., *Man's Rise to Civilization as Shown by the Indians of North America from Primeval Times to the Coming of the Industrial State*, E. P. Dutton & Co., New York, 1968.

6 Turner, F. J., *The Frontier in American History*, Henry Holt & Co., New York, 1893.
  Goldstein, R. A., *French–Iroquois Diplomatic and Military Relations 1609–1701*, Mouton, Paris, 1969.

7 Smith, J., *The True Travels, Adventures and Observations of Captain John Smith in Europe, Asia, Africa and America, and the General History of Virginia, New England*

and the Summer Isles, Books 1–3, Cambridge University Press, Cambridge, 1608–30, reprinted 1908.

8 Lurie, E., *Louis Agassiz: A Life in Science*, Johns Hopkins University Press, Baltimore, 1988.

TWENTY-ONE: ENGLISH COLONIES ALL

1 Shenkman, R., *Legends, Lies and Cherished Myths of American History*, Harper-Perennial, New York, 1989.

2 Bradford, W., *Of Plymouth Plantation 1620–1647*, Modern Library College Editions, New York, 1952. First pub. as *History of Plymouth Plantation* in 1856.

3 Sauer, C. O., *Sixteenth Century North America*, University of California Press, Berkeley, 1975.

4–6 Bradford, W., *Of Plymouth Plantation*, op. cit.

7 Boorstin, D. J., *The Americans: The Colonial Experience*, Cardinal, London, 1958, reprinted 1991.

8 & 9 Bradford, W., *Of Plymouth Plantation*, op. cit.

10 & 11 Boorstin, D. J., *The Americans*, op. cit.

12 Turner, F. J., *The Frontier in American History*, Henry Holt & Co., New York, 1893.

13 Wilson, J., *The Earth Shall Weep: A History of Native America*, Atlantic Monthly Press, New York, 1999.

14 Shenkman, R., *Legends, Lies and Cherished Myths of American History*, op. cit.

15 Withington, A. F., *Towards a More Perfect Union: Virtue and the Formation of American Republics*, Oxford University Press, New York, 1991.

16 ibid.

Trollope, F., *Domestic Manners of the Americans*, Alan Sutton, Dover, New Hampshire, 1832.

17 & 18 Withington, A. F., *Towards a More Perfect Union*, op. cit.

19 Matthiessen, P., *Wildlife in America*, Penguin Nature Classics, Penguin Books, Harmondsworth, 1959.

20 ibid.

Hampton, B., *The Great American Wolf*, Owl Books, New York, 1997.

TWENTY-TWO: CONCEIVED IN LIBERTY

1 Degler, C. N., *Out of Our Past: The Forces That Shaped Modern America*, Harper & Row, New York, 1959.

2 Codrescu, A., 'Hidden History in the "Broad Shouldered" City of Chicago' in *Spirit*, Southwestern Airlines, April–May 1999, pp. 63–64, 128, 131–37, 143.

3 Frazier, I., *Great Plains*, Farrar, Straus & Giroux, New York, 1989.

4 & 5 Turner, F. J., *The Frontier in American History*, first pub. Henry Holt & Co, New York, 1893, reprinted 1947.

6 Boorstin, D. J., *The Americans: The Colonial Experience*, Cardinal, London, 1958, reprinted 1991.
  Pyne, S. J., *Fire in America: A Cultural History of Wildland and Rural Fire*, University of Washington Press, Seattle, 1982.

7 Shenkman, R., *Legends, Lies and Cherished Myths of American History*, Harper-Perennial, New York, 1989.
  Turner, F. J., *The Frontier in American History*, first pub. Henry Holt & Co., New York, 1893.

8–11 Lever, C., *Naturalised Mammals of the World*, Longman, London, 1985.

TWENTY-THREE: THE FATAL IMPACT

1 Trollope, F., *Domestic Manners of the Americans*, Alan Sutton, Dover, New Hampshire, 1832.

2 Sauer, C. O., *Sixteenth Century North America*, University of California Press, Berkeley, 1975.

3 Casey, J. M. & Rogers, R. A., 'New Extinction of a Large Widely Distributed Fish' in *Science* vol. 281, 1998, pp. 690–92.

4 Matthiessen, P., *Wildlife in America*, Penguin Nature Classics, Penguin Books, Harmondsworth, 1959.

5 & 6 Bradford, W., *Of Plymouth Plantation 1620–1647*, Modern Library College Editions, New York, 1952. First pub. as *History of Plymouth Plantation* in 1856.

7 Catlin, G., *Letters and Notes on the Manners, Customs and Conditions of North American Indians*, Dover Publications, New York, 1844, reprinted 1973.
  Frazier, I., *Great Plains*, Farrar, Straus & Giroux, New York, 1989.

8 ibid.

9 Fiedel, S. J., *Prehistory of the Americas*, Cambridge University Press, Cambridge, 2nd edn 1993.

10 Henneberg, M. & Henneberg, R. J., 'Treponematosis in an Ancient Greek Colony of Metaponto, Southern Italy, 580–250 BC' in *La Syphilis avant 1493 en Europe?* O. Dutour, G. Pálfi, J. Berato & J-P. Brun eds, Actes du Colloque de Tulon, Centre Archéologique du Var and Paris, Editions Errance, 1994, pp. 92–98.
  Steyn, M. & Henneberg, M., 'Pre-Columbian Presence of Treponemal Disease: A Possible Case from Iron Age Southern Africa' in *Current Anthropology* vol. 36, 1995, pp. 870–73.

11 Farb, P., *Man's Rise to Civilization as Shown by the Indians of North America from Primeval Times to the Coming of the Industrial State*, E. P. Dutton & Co., New York, 1968.

12 'Mourt's Relation', (facsimile) Applewood Books, Bedford, 1622.

13 Bradford, W., *Of Plymouth Plantation*, op. cit.
    Wilson, J., *The Earth Shall Weep: A History of Native America*, Atlantic Monthly Press, New York, 1999.

14 Crosby, A. W., *The Columbian Exchange: Biological and Cultural Consequences of 1492*, Greenwood Press, Connecticut, 1972.

15 Turner, F. J., *The Frontier in American History*, first pub. Henry Holt & Co., New York, 1893, reprinted 1947.

16 Wilson, J., *The Earth Shall Weep*, op. cit.

17 Trollope, F., *Domestic Manners of the Americans*, op. cit.

18 Wilson, J., *The Earth Shall Weep*, op. cit.

19 Farb, P., *Man's Rise to Civilization...*, op. cit.
    Nye, R. B. & Morpurgo, J. E., *A History of the United States, Volume 2: The Growth of the USA*, Penguin Books, Harmondsworth, 1970.

20 Wilson, J., *The Earth Shall Weep*, op. cit.

TWENTY-FOUR: AMERICA UNDER THE GUN

1 Stephan, S. A., *The Passing of the Passenger Pigeon*, Game Stories no. 18–19, 1932.

2 & 3 Matthiessen, P., *Wildlife in America*, Penguin Nature Classics, Penguin Books, Harmondsworth, 1959.

4 Gollup, J. B., Barry, T. W. & Iversen, E. H., *Eskimo Curlew: A Vanishing Species?*, special publication no. 17, Saskatchewan Natural History Society, 1986.
   Matthiessen, P., *Wildlife in America*, op. cit.

5 ibid.

6 Teale, E. W., *Autumn across America*, St Martin's Press, New York, 1956.

7 Clutton-Brock, J., *Horse Power*, Natural History Museum Publications, London, 1992.

8–10 Hornaday, W. T., 'The Extermination of the American Bison', *Smithsonian Report*, Washington, 1889, pp. 369–548.

11 Doughty, R., *Wildlife and Man in Texas: Environmental Change and Conservation*, Texas A&M University Press, Texas, 1983.
   Hampton, B., *The Great American Wolf*, Owl Books, New York, 1997.

12 Knapp, A. K., Blair, J. M., Briggs, J. M., Collins, S. L., Hartnett, D. C., Johnson, L. C. & Towne, E. G., 'The Keystone Role of Bison in North American Tallgrass Prairie' in *Bioscience* vol. 49, 1999, pp. 39–50.

13 Pringle, H., 'The Slow Birth of Agriculture' in *Science* vol. 282, 20 Nov. 1998, pp. 1446–50.

14 Hampton, B., *The Great American Wolf*, op. cit.

15 Hornaday, W. T., op. cit.

16 & 17 Hampton, B., *The Great American Wolf*, op. cit.

18 Hornaday, W. T., op. cit.

19 Hampton, B., *The Great American Wolf*, op. cit.

20 Davis, L. B., 'The Twentieth Century Commercial Mining of Northern Plains Bison Kills' in *Bison Procurement and Utilisation: A Symposium*, L. B. Davis & M. Wilson eds, memoir 14, *Plains Anthropologist*, 1978, pp. 254–85.

21 Hampton, B., *The Great American Wolf*, op. cit.

TWENTY-FIVE: THE MAKING OF A GIANT

1 Boorstin, D. J., *The Genius of American Politics*, University of Chicago Press, Chicago, 1953.

2 Bruchey, S., *The Roots of American Economic Growth 1607–1861*, Harper & Row, New York, 1965.

3–5 Hounshell, D. A., *From the American System to Mass Production, 1800–1932*, Johns Hopkins University Press, Baltimore, 1984.

6 Schlereth, T. J., *Victorian America: Transformations in Everyday Life*, Harper-Perennial, New York, 1991.

7 Bailyn, B., *The Peopling of British North America: An Introduction*, Vintage Books, Wisconsin, 1986, reprinted 1988.

8 Frazier, I., *Great Plains*, Farrar, Straus & Giroux, New York, 1989.

9 Ehrlich, P. R., Dobkin, D. S. & Wheye, D., *The Birder's Handbook*, Simon & Schuster, New York, 1988.
   Matthiessen, P., *Wildlife in America*, Penguin Nature Classics, Penguin Books, Harmondsworth, 1959.

10 ibid.

11 & 12 Reisner, M., *Cadillac Desert: The American West and Its Disappearing Water*, Viking, New York, 1986.

13–15 Wilcove, D. S., *The Condor's Shadow: The Loss and Recovery of Wildlife in America*, Freeman & Co., New York, 1999.

16 Frazier, I., *Great Plains*, op. cit.
   Reisner, M., *Cadillac Desert*, op. cit.

17 Wilcove, D. S., *The Condor's Shadow*, op. cit.

18 & 19 Reisner, M., *Cadillac Desert*, op. cit.

TWENTY-SIX: WHAT'S HOME ON THE RANGE?

1 Krausman, P. R., 'Conflicting Views of Ungulate Management in North America's Western National Parks', *Wildlife Society Bulletin* vol. 26, pp. 369–71, 1999.

2 Boyce, M. S., *The Jackson Elk Herd: Intensive Wildlife Management in North America*, Cambridge University Press, Cambridge, 1989.

3 Fiedel, S. J., *Prehistory of the Americas*, Cambridge University Press, Cambridge, 2nd edn, 1993.

4 Kay, C. E., American Anthropological Association Abstracts, 1993.

5 & 6 Krech III, S., *The Ecological Indian: Myth and History*, Norton & Co., New York, 1999.

7 Kay, C. E., 'Too Many Elk in Yellowstone?', *Western Wildlands*, Fall 1987.

8 Campbell, C. C., Campbell, I. D., Blyth, C. B. & McAndrews, J. H., 'Bison Extirpation May Have Caused Aspen Expansion in Western Canada' in *Ecography* vol. 17, 1994, pp. 360–62.

9 Brandner, T. A., Peterson, R. O. & Risenhoover, K. L., 'Balsam Fir on Isle Royale: Effects of Moose Herbivory and Population Density' in *Ecology* vol. 71, 1990, pp. 155–64.

Kay, C. E., 'Reduction of Willow Seed Production by Ungulate Browsing in Yellowstone National Park' in *Symposium on Ecology and Management of Riparian Shrub Communities*, Intermontane Research Station Forest Service, US Dept. Ag. Ogden, Utah, 1992, pp. 92–99.

Kay, C. E., 'Effects of Browsing by Native Ungulates on Shrub Growth and Seed Production in the Greater Yellowstone Ecosystem: Implications for Revegetation, Restoration, and "Natural Regulation" Management', paper presented at symposium on wildland shrub & arid land restoration, Las Vegas, Nov. 1993.

Kay, C. E. & Wagner, F. H., 'Historic Condition of Woody Vegetation: Plants and Their Environments', scientific conference on the Greater Yellowstone ecosystem (1st), US Dept. Interior, *National Parks Service Technical Report 93*, Natural Resources Publication Office, Denver, 1994.

10 Owen-Smith, R. N., *Megaherbivores: The Influence of Very Large Body Size on Ecology*, Cambridge University Press, 1988.

11 Fiedel, S. J., *Prehistory of the Americas*, op. cit.

12 Wilson, J., *The Earth Shall Weep: A History of Native America*, Atlantic Monthly Press, New York, 1999.

13 Mellink, E., 'Use of Sonoran Rangelands: Lessons from the Pleistocene', chapter 4 in *Late Quaternary Environments and Deep History: A Tribute to Paul S. Martin*, D. W. Steadman & J. I. Mead eds, Mammoth Site Hot Springs, scientific papers vol. 3, 1995.

14 Klippel, W. E. & Parmalee, P. W., 'Armadillos in North American Late Pleistocene Contexts', *Contributions in Quaternary Vertebrate Paleontology: A Volume in Memorial to John E. Guilday*, special publication of Carnegie Museum of Natural History, Pittsburgh, 1984.

TWENTY-SEVEN: REINVENTING AMERICA

1 Boorstin, D. J., *The Americans: The Colonial Experience*, Cardinal, London, 1958, reprinted 1991.

  Reisner, M., *Cadillac Desert: The American West and Its Disappearing Water*, Viking, New York, 1986.

2 ibid.

3 Keynes, J. M., *The General Theory of Employment, Interest and Money*, Harcourt Brace, San Diego, 1935.

4 Turner, F. J., *The Frontier in American History*, first pub. Henry Holt & Co., New York, 1893, reprinted 1947.

BIBLIOGRAPHY

Agenbroad, L. D., Mead, J. I. & Nelson, L. W. eds, *Megafauna and Man: Discovery of America's Heartland*, Mammoth Site of Hot Springs, South Dakota, Inc. Scientific Papers vol. 1, 1990.

Archibald, J. D., 'A Study of Mammalia and Geology across the Cretaceous-Tertiary Boundary in Garfield County, Montana' in *University of California Publications in Geological Sciences* vol. 122, University of California Press, California, 1982.

Archibald, J. D., 'Fossil Evidence for a Late Cretaceous Origin of "Hoofed" Mammals' in *Science* vol. 272, 1996.

Axelrod, D. I., 'Age and Origin of Sonoran Desert Vegetation', *California Academy of Science Occasional Paper 132*, 1979.

Axelrod, D. I., 'Contributions to the Neogene Paleobotany of Central California', *University of California Publications in Geological Sciences* vol. 121, University of California Press, California, 1980.

Bailyn, B., *The Peopling of British North America: An Introduction*, Vintage Books, Wisconsin, 1986, reprinted 1988.

Barber, L., *The Heyday of Natural History*, Jonathan Cape, London, 1980.

Barnosky, A. D., 'Age of the Mid-Tertiary Unconformity in the Western Rocky Mountains and Miocene Biogeography', Society of Vertebrate Paleontology Abstracts, 1998.

Beard, K. C., 'East of Eden: Asia as an Important Center of Taxonomic Origination in Mammalian Evolution' in *Dawn of the Age of Mammals in Asia*, K. C. Beard & M. R. Dawson eds, *Bulletin of the Carnegie Museum of Natural History*, vol. 34, Pittsburgh, 1998.

Behler, J. L. & King, F. W., *National Audubon Society Field Guide to North American Reptiles and Amphibians*, 16th edn, Alfred A. Knopf, New York, 1998.

Benson, R. D., '*Presbyornis Isoni* and Other Late Paleocene Birds from North Dakota' in *Smithsonian Contribution to Palaeobiology* vol. 89, 2000.

Betts, C., 'The Yale College Expedition of 1870' in *Harper's New Monthly Magazine*, Oct. 1871.

Black, C. C., 'Holarctic Evolution and Dispersal of Squirrels (Rodentia: Sciuridae)' in *Evolutionary Biology* vol. 6, T. Dobshansky, M. K. Hecht & W. C. Steere eds, Appleton-Century-Crofts, New York, 1972.

Boorstin, D. J., *The Americans: The Colonial Experience*, Cardinal, London, 1958, reprinted 1991.

Boorstin, D. J., *The Genius of American Politics*, University of Chicago Press, Chicago, 1953.

Bourgeois, J., Hansen, T. A., Wiberg, P. L. & Kaufman, E. G., 'A Tsunami Deposit at the Cretaceous-Tertiary Boundary in Texas' in *Science* vol. 241, 1988.

Boyce, M. S., *The Jackson Elk Herd: Intensive Wildlife Management in North America*, Cambridge University Press, Cambridge, 1989.

Bradford, W., *Of Plymouth Plantation 1620–1647*, Modern Library College Editions, New York, 1952. First pub. as *History of Plymouth Plantation* in 1856.

Brandner, T. A., Peterson, R. O. & Risenhoover, K. L., 'Balsam Fir on Isle Royale: Effects of Moose Herbivory and Population Density' in *Ecology* vol. 71, 1990.

Brodkorb, P., 'A Giant Flightless Bird from the Pleistocene of Florida' in *The Auk* vol. 8, American Ornithologists' Union, 1963.

Bruchey, S., *The Roots of American Economic Growth 1607–1861*, Harper & Row, New York, 1965.

Bryant, L. J., 'Non Dinosaurian Lower Vertebrates across the Cretaceous-Tertiary Boundary in Northeastern Montana' in *University of California Publications in Geological Sciences* vol. 134, University of California Press, California, 1989.

Buchardt, B., 'Oxygen Isotope Paleotemperatures from the Tertiary Period in the North Sea Area' in *Nature* vol. 275, 1979.

Burnham, R. T., Wing, S. L. & Parker, G. G., 'The Reflection of Deciduous Forest Communities in Leaf Litter: Implications for Autochthonous Litter Assemblages from the Fossil Record' in *Paleobiology* vol. 18, 1992.

Byers, J. A., *The American Pronghorn: Social Adaptations and the Ghosts of Predators Past*, University of Chicago Press, Chicago, 1997.

Callahan, J. E., 'Velocity Structure and Flux of the Antarctic Circumpolar Current' in *Journal of Geophysical Research* vol. 76, 1971.

Campbell, C. C., Campbell, I. D., Blyth, C. B. & McAndrews, J. H., 'Bison Extirpation May Have Caused Aspen Expansion in Western Canada' in *Ecography* vol. 17, 1994.

Campbell, K. E. & Tonni, E. P., 'Preliminary Observations on the Paleobiology and Evolution of Teratorns (Aves: Teratornithidae)' in *Journal of Vertebrate Paleontology* vol. 1, 1981.

Carleton, M. D. & Olson, S. L., 'Amerigo Vespucci and the Rat of Fernando de Noronha: A New Genus and Species of Rodentia (Muridae: Sigmodontinae) from a Volcanic Island off Brazil's Continental Shelf' in *American Museum Natural History Novitates 3256*, 1999.

Casey, J. M. & Rogers, R. A., 'New Extinction of a Large Widely Distributed Fish' in *Science* vol. 281, 1998.

Catlin, G., *Letters and Notes on the Manners, Customs and Conditions of North American Indians*, Dover Publications, New York, 1844, reprinted 1973.

Censky, E. J., Hodge, K. & Dudley, J., 'Over-Water Dispersal of Lizards Due to Hurricanes' in *Nature* vol. 395, 1998.

Chace, A., *Playing God in Yellowstone: The Destruction of America's First National Park*, First Harvest/HBT, San Diego, 1987.

Cifelli, R. L., 'The Origin and Affinities of the South American Condylarthra and Early Tertiary Litopterna (Mammalia)' in *American Museum Natural History Novitates 2772*, 1983.

Clutton-Brock, J., *Horse Power*, Natural History Museum Publications, London, 1992.

Clyde, W. C. 'Effects of Latest Paleocene Climate Change and Immigration on Local Mammalian Communities in the Bighorn Basin, Wyoming: New Evidence from the McCulloch Peaks', Society of Vertebrate Paleontology Abstracts, 1997.

Codrescu, A., 'Hidden History in the "Broad Shouldered" City of Chicago' in *Spirit*, Southwestern Airlines, April–May 1999.

COHMAP members, 'Climatic Changes of the Last 18,000 Years: Observations and Model Simulations' in *Science* vol. 241, 1988.

Colbert, E. H., *The Great Dinosaur Hunters and Their Discoveries*, Dover, New York, 1984.

Contributions in Quaternary Vertebrate Paleontology: A Volume in Memorial to John E. Guilday, Carnegie Museum of Natural History, Pittsburgh, 1984.

Crosby, A. W., *The Columbian Exchange: Biological and Cultural Consequences of 1492*, Greenwood Press, Connecticut, 1972.

Darlington, P. J., *Zoogeography: The Geographical Distribution of Animals*, John Wiley & Sons, New York, 1957.

Darwin, C., *Journal of Researches into the Geology and Natural History of the Various Countries Visited by H.M.S. Beagle*, (facsimile) Hafner, London, 1839.

Davis, L. B., 'The Twentieth Century Commercial Mining of Northern Plains Bison Kills' in *Bison Procurement and Utilisation: A Symposium*, L. B. Davis & M. Wilson eds, *Plains Anthropologist* memoir 14, 1978.

Degler, C. N., *Out of Our Past: The Forces That Shaped Modern America*, Harper & Row, New York, 1959.

De Tocqueville, A., *Democracy in America*, trans. G. Lawrence, J. P. Mayer & M. Lerner eds, Harper & Row, New York, 1966.

D'Hondt, S., Pilson, M. E. Q., Sigurdsson, H., Hanson A. K. Jr, & Carey, S., 'Surface-Water Acidification and Extinction at the Cretaceous-Tertiary Boundary' in *Geology* vol. 22, 1994.

Diamond, J., *Guns, Germs and Steel: A Short History of Everybody for the Last 13,000 Years*, Jonathan Cape, London, 1997.

Diaz del Castillo, B., *The Discovery and Conquest of Mexico, 1517–1521*, trans. A. P. Maudslay, Grove Press, New York, 1958.

Dickens, C., *American Notes for General Circulation*, Chapman & Hall, London, 1842, reprinted Penguin, Harmondsworth, 1972.

Dillehay, T. D., *Monte Verde: A Late Pleistocene Settlement in Chile*, vol. 2, Smithsonian Institution Press, Washington, 1997.

Doughty, R., *Wildlife and Man in Texas: Environmental Change and Conservation*, A&M University Press, Texas, 1983.

Draper, T., *A Struggle for Power*, Times Books, New York, 1996.

Earle, A. M., *Curious Punishments of Bygone Days*, Herbert S. Stone & Co., Chicago, 1896, reprinted by Applewood Books, Bedford, 1995.

Ehrlich, P. R., Dobkin, D. S. & Wheye, D., *The Birder's Handbook*, Simon & Schuster, New York, 1988.

Elias, S. A., Short, S. K. & Birks, H. H., 'Late Wisconsin Environments on the Bering Land Bridge' in *Palaeogeography, Palaeoclimatology, Palaeoecology* vol. 136, 1997.

Eliot, C. W., *American Historical Documents 1000–1904*, Harvard Classics, F. P. Collier & Son, New York, 1938.

Emslie, S. D., 'The Fossil History and Phylogeneric Relationships of Condors (Ciconiformes: Vulturidae) in the New World' in *Journal of Vertebrate Paleontology* vol. 8, Mammoth Site Hot Springs, 1988.

Esat, T. M., McCulloch, M. T., Chappell, J., Pillans, B. & Omura, A., 'Rapid Fluctuations in Sea Level Recorded at Huon Peninsula during the Penultimate Glaciation' in *Science* vol. 283, 1999.

Farb, P., *Man's Rise to Civilization as Shown by the Indians of North America from Primeval Times to the Coming of the Industrial State*, E. P. Dutton & Co., New York, 1968.

Fiedel, S. J., *Prehistory of the Americas*, Cambridge University Press, Cambridge, 2nd edn, 1993.

Fisher, D. C. 'Experiments on Subaqueous Meat Caching' in *CPR* vol. 12, 1995.

Fisher, D. C., 'Testing Late Pleistocene Extinction Mechanisms with Data on Mastodon and Mammoth Life History', Society of Vertebrate Paleontology Abstracts vol. 16 (3), 1996.

Fisher, D. C., 'Mastodon Procurement by Paleoindians of the Great Lakes Region: Hunting or Scavenging?' in *The Evolution of Human Hunting*, M. H. Nitecki & D. V. Nitecki eds, Plenum Press, New York, 1987.

Fisher D. C., 'Extinctions of Proboscideans in North America', chapter 30 in *The*

Proboscidea, J. Shoshani & P. Tassy eds, Museum of Paleontology, Michigan, 1997.

Flannery, T. F., *The Future Eaters: An Ecological History of the Australasian Lands and People*, Reed Books, Port Melbourne, 1994.

Flint, R. F., *Glacial and Quaternary Geology*, John Wiley & Sons, New York, 1971.

Flynn, L. J., Tedford, R. H. & Zhanxiang, Q., 'Enrichment and Stability in the Pliocene Mammalian Fauna of North China' in *Paleobiology* vol. 17, 1991.

Foss, S. E., 'Behavioural Interpretation of the Entelodontidae (Mammalia: Artiodactyla)', Society of Vertebrate Paleontology Abstracts, 1998.

Fox, J. W., Smith, C. B. & Williams, K. T. eds, *Proboscidean and Paleoindian Interactions*, Baylor University Press, Waco, 1992.

Frakes, L. A., *Climates throughout Geological Time*, Elsevier, New York, 1979.

Franklin, B., 'Interest of Great Britain Considered with Regard to Her Colonies and the Acquisitions of Canada & Guadaloupe' (cited in Draper 1996:11), 1760.

Frazier, I., *Great Plains*, Farrar, Straus & Giroux, New York, 1989.

Gamble, C., *The Palaeolithic Settlement of Europe*, Cambridge University Press, London, 1986.

Gittleman, J. L. ed., *Carnivore Behaviour, Ecology and Evolution*, Cornell University Press, Ithaca, 1989.

Goldstein, R. A., *French-Iroquois Diplomatic and Military Relations 1609–1701*, Mouton, Paris, 1969.

Gollup, J. B., Barry, T. W. & Iversen, E. H., *Eskimo Curlew: A Vanishing Species?*, special pub. no. 17, Saskatchewan Natural History Society, 1986.

Graham, R. W., Farlow, J. O. & Vandike, J. E., 'Tracking Ice Age Fields: Identification of Tracks of *Panthera atrox* from a Cave in Southern Missouri, USA in *Palaeoecology and Paleoenvironments of Late Cenozoic Mammals*, K. M. Stewart & K. L. Seymour eds, University of Toronto Press, Toronto, 1996.

Grayson, D. K., 'The Chronology of North American Late Pleistocene Extinctions' in *Journal of Archaeological Science* vol. 16, 1989.

Guthrie, R. D., 'Alaskan Megabucks, Megabulls and Megarams: The Issue of Pleistocene Gigantism', *Contributions in Quaternary Vertebrate Paleontology: A Volume in Memorial to John E. Guilday*, Carnegie Museum of Natural History, Pittsburgh, 1984.

Guthrie, R. D., 'Bison Evolution and Zoogeography in North America during the Pleistocene' in *Quarterly Review of Biology* vol. 45, 1970.

Guthrie, R. D., *Frozen Fauna of the Mammoth Steppe: The Story of Blue Babe*, University of Chicago Press, Chicago, 1990.

Hampton, B., *The Great American Wolf*, Owl Books, New York, 1997.

Hanson, T. A., Farrell, B. R. & Banks Upshaw III, 'The First Two Million Years after the Cretaceous-Tertiary Boundary in East Texas: Rate and Paleoecology of the Molluscan Recovery' in *Paleobiology* vol. 19, 1993.

Harris, A. H., *Late Pleistocene Vertebrate Paleoecology of the West*, University of Texas Press, Austin, 1985.

Haynes, C. V., 'Elephant Hunting in North America' in *Scientific American* vol. 214 (6), 1966.

Heizer, R. F. & Whipple, M. A., *The California Indians: A Source Book*, University of California Press, Berkeley, 1951.

Henneberg, M. & Henneberg, R. J., 'Treponematosis in an Ancient Greek Colony of Metaponto, Southern Italy, 580–250 BC' in *La Syphilis avant 1493 en Europe?*, O. Dutour, G. Pálfi, J. Berato & J-P. Brun eds, Actes du Colloque de Tulon, Centre Archéologique du Var and Paris, Editions Errance, 1994.

Holman, J. A., *Pleistocene Amphibians and Reptiles in North America*, Oxford University Press, Cary, 1995.

Holman, J. A. & Clausen, C. J., 'Fossil Vertebrates Associated with Paleoindian Artifact at Little Salt Spring, Florida' in *Journal of Vertebrate Paleontology* vol. 4, 1984.

Holman, J. A., '*Texasophis galbraithi* New Species: The Earliest New World Colubrid Snake' in *Journal of Vertebrate Paleontology* vol. 3, 1983.

Hope, S., 'A New Species of *Graculavus* from the Cretaceous of Wyoming' in *Smithsonian Contribution to Palaeobiology* vol. 89, 2000.

Hornaday, W. T., 'The Extermination of the American Bison', *Smithsonian Report*, Washington, 1889.

Hounshell, D. A. *From the American System to Mass Production, 1800–1932*, Johns Hopkins University Press, Baltimore, 1984.

Hulbert, R. C., 'Late Neogene *Neohipparion* (Mammalia: Equidae) from the Gulf Coastal Plain of Florida and Texas' in *Journal of Paleontology* vol. 61, 1987.

Iturralde-Vinent, M. A. & MacPhee, R. D. E., 'Paleogeography of the Caribbean Region: Implications for Cenozoic Biogeography' in *Bulletin of the American Museum of Natural History* vol. 238, 1999.

Janis, C. M., 'Tertiary Mammal Evolution in the Context of Changing Climates, Vegetation, and Tectonic Events' in *Annual Review of Ecological Systematics* vol. 24, 1993.

Janis, C. M., Scott, K. M. & Jacobs, L. L. eds, *Evolution of Tertiary Mammals of North America, Volume 1: Terrestrial Carnivores, Ungulates, and Ungulate-Like Mammals*, Cambridge University Press, Cambridge, 1998.

Johnson, K. R., 'Leaf-Fossil Evidence for Extensive Floral Extinction at the

Cretaceous-Tertiary Boundary, North Dakota, USA' in *Cretaceous Research* vol. 13, 1992.

Johnson, K. R., Nichols, D. J., Attrep, M. & Orth, C. J., 'High-Resolution Leaf-Fossil Record Spanning the Cretaceous-Tertiary Boundary' in *Nature* vol. 340, 1989.

Johnston, P. A. & Haggart, J. W. eds, *Bivalves: An Eon of Evolution*, University of Calgary Press, Calgary, 1988.

Jones, R., 'Folsom and Talgai: Cowboy Archaeology in Two Continents', chapter 2 in *Approaching Australia: Papers from the Harvard Australian Studies Symposium*, H. Bolitho & C. Wallace-Crabbe eds, Harvard University Committee on Australian Studies, Cambridge, 1998.

Kahlke, R. D., 'The History of the Origin, Evolution and Dispersal of the Late Pleistocene *Mammuthus-Coelodonta* Faunal Complex in Eurasia (Large Mammals)', Mammoth Site Hot Springs, South Dakota, scientific papers, 1999.

Kay, C. E., 'Too Many Elk in Yellowstone?', *Western Wildlands*, Fall 1987.

Kay, C. E., 'Reduction of Willow Seed Production by Ungulate Browsing in Yellowstone National Park' in *Symposium on Ecology and Management of Riparian Shrub Communities*, Intermontane Research Station Forest Service, US Dept. Agriculture, Utah, 1992.

Kay, C. E., 'Effects of Browsing by Native Ungulates on Shrub Growth and Seed Production in the Greater Yellowstone Ecosystem: Implications for Revegetation, Restoration, and "Natural Regulation" Management', paper presented at symposium on wildland shrub & arid land restoration, Las Vegas, Nov. 1993.

Kay, C. E. & Wagner, F. H., 'Historic Condition of Woody Vegetation: Plants and Their Environments', scientific conference on the Greater Yellowstone ecosystem (1st), US Dept. of the Interior, *National Parks Service Technical Report 93*, Natural Resources Publication Office, Denver, 1994.

Keynes, J. M., *The General Theory of Employment, Interest and Money*, Harcourt Brace, San Diego, 1935.

Knapp, A. K., Blair, J. M., Briggs, J. M., Collins, S. L., Hartnett, D. C., Johnson, L. C. & Towne, E. G., 'The Keystone Role of Bison in North American Tallgrass Prairie' in *Bioscience* vol. 49, 1999.

Kontrimavichus, V. L. ed., *Beringia in the Cenozoic Era*, Oxonian Press, New Delhi, 1984.

Krausman, P. R., 'Conflicting Views of Ungulate Management in North America's Western National Parks' in *Wildlife Society Bulletin* vol. 26, 1999.

Krech III, S., *The Ecological Indian: Myth and History*, W. W. Norton & Co., New York, 1999.

Kurlansky, M., *Cod: A Biography of the Fish That Changed the World*, Jonathan Cape, London, 1997.

Kurten, B. & Anderson, E., *Pleistocene Mammals of North America*, Columbia University Press, New York, 1980.

Kyte, F. T. 'A Meteorite from the Cretaceous-Tertiary Boundary' in *Nature* vol. 396, 1998.

Laughlin, W. S., 'Eskimos and Aleuts: Their Origins and Evolution' in *Science* vol. 142, 1963.

Leidy, J., 'On Some New Species of Fossil Mammalia from Wyoming' in *Proceedings of the Academy of Natural Sciences of Philadelphia*, 1872.

Leonard, J. A., Wayne, R. K. & Cooper, A., 'Population Genetics of Ice Age Brown Bears', *PNAS 97*, 1999.

Lepper, B. T., Frolking, T. A., Fisher, D. C., Goldstein, G., Sanger, J. E., Wymer, D. A., Ogden III, J. G. & Hooge, P. E., 'Intestinal Contents of a Late Pleistocene Mastodon from Midcontinental North America' in *Quaternary Research* vol. 36, 1991.

Lever, C., *Naturalised Mammals of the World*, Longman, London, 1985.

Lewin, R., *Patterns in Evolution: The New Molecular View*, Scientific American Library, New York, 1999.

Li, H-L., '*Metasequoia*: A Living Fossil' in *American Scientist* vol. 52, 1964.

Lillegraven, J. A., 'Evolution of Wyoming's Paleocene Eastern Lowlands Related to Tectonic Loading by the Cannonball Sea', Society of Vertebrate Paleontology Abstracts, 1997.

Lillegraven, J. A. & Eberle, J. J., 'Vertebrate Faunal Changes through Lancian and Puercan Time in Southern Wyoming' in *Journal of Paleontology* vol. 73 (4), 1999.

Lofgren, D. L., 'The Bug Creek Problem and the Cretaceous-Tertiary Transition at McGuire Creek, Montana', *University of California Publications in Geological Sciences* vol. 40, University of California Press, California, 1995.

Lurie, E., *Louis Agassiz: A Life in Science*, Johns Hopkins University Press, Baltimore, 1988.

Maas, M. C. & Krause, D. W., 'Mammalian Turnover and Community Structure in the Paleocene of North America' in *Historical Biology* vol. 8, 1994.

MacFadden, B. J., *Fossil Horses: Systematics, Paleobiology and Evolution of the Family Equidae*, Cambridge University Press, Cambridge, 1992.

MacLeod, N. & Keller, G. eds, *Cretaceous-Tertiary Mass Extinctions: Biotic and Environmental Changes*, W. W. Norton & Co., New York, 1996.

MacPhee, R. D. E., Flemming, C. & Lunde, D. P., '"Last Occurrence" of the Antillean Insectivoran *Nesophontes*: New Radiometric Dates and Their Interpretation' in *American Museum Natural History Novitates 3261*, 1999.

Madden, C. T. & Dalquest, W. W., 'The Last Rhinoceros in North America' in *Journal of Vertebrate Paleontology* vol. 10, 1990.

Marsh, G. H. & Laughlin, W. S., 'Human Anatomical Knowledge among the Aleutian Islanders' in *Southwestern Journal of Anthropology* vol. 12, 1956.

Marsh, O. C., 'On the Gigantic Fossil Mammals of the Order Dinocerata', *American Journal of Science* vol. 3, 1873.

Marsh, O. C., 'Dinocerata: A Monograph of an Extinct Order of Gigantic Mammals', *Monograph of the United States Geological Survey* vol. 10, 1886.

Martin, P. S., 'Vanishings, and Future, of the Prairie' in *Geoscience and Man* vol. 10, 1975.

Martin, P. S. and Klein, R. G. eds, *Quaternary Extinctions: A Prehistoric Revolution*, University of Arizona Press, Tucson, 1984.

Martin, R. A. & Barnosky, A. D. eds, *Morphological Change in Quaternary Mammals of North America*, Cambridge University Press, New York, 1993.

Marton, G. & Buffler, R. T., 'Application of Simple Shear Model to the Evolution of Passive Continental Margins in the Gulf of Mexico Basin' in *Geology* vol. 21, 1993.

Matthiessen, P., *Wildlife in America*, Penguin Nature Classics, Penguin, Harmondsworth, 1959.

McCrum, R., Cran, W. & MacNeil, R., *The Story of English*, Faber & Faber, London, 1987.

McKenna, M., 'Was Europe Connected Directly to North America Prior to the Mid-Eocene?' in *Evolutionary Biology* vol. 6, 1972.

Melosh, H. J., Schneider, N. M., Zahnle, K. J. & Latham, D., 'Ignition of Global Wildfires at the Cretaceous-Tertiary Boundary' in *Nature* vol. 343, 1990.

Meltzer, D. J. & Mead, J. I., 'The Timing of Late Pleistocene Mammalian Extinctions in North America' in *Quaternary Research* vol. 19, 1983.

Miller, G. H., Magee, J. W., Johnson, B. J., Fogel, M. L., Spooner, N. A., McCulloch, M. T. & Ayliffe, L. K., 'Pleistocene Extinction of *Genyornis newtoni*: Human Impact on Australian Megafauna' in *Science* vol. 283, 1999.

Morey, D. F. & Wiant, M. D., 'Early Holocene Dog Burials from the North American Midwest' in *Current Anthropology* vol. 33, 1992.

'Mourt's Relation', (facsimile) Applewood Books, Bedford, 1622.

Nye, R. B. & Morpurgo, J. E., *A History of the United States: The Growth of the USA*, vol. 2, Penguin, Harmondsworth, 1970.

Obradovich, J. D., 'A Cretaceous Time Scale' in *Evolution of the Western Interior Basin*, W. G. E. Caldwell & E. G. Kauffman eds, Geological Association of Canada, special paper no. 39, 1993.

Oksanen, L., 'Ecosystem Organisation: Mutualism and Cybernetics or Plain Darwinian Struggle for Existence?' in *The American Naturalist* vol. 131, 1988.

Owen-Smith, R. N., *Megaherbivores: The Influence of Very Large Body Size on Ecology*, Cambridge University Press, Cambridge, 1988.

Owen-Smith, R. N., 'Pleistocene Extinctions: The Pivotal Role of Megaherbivores' in *Paleobiology* vol. 13, 1987.

Owen-Smith, R. N. & Danckwerts, J. E., 'Herbivory', chapter 17 in *The Vegetation of Southern Africa*, Cambridge University Press, Cambridge, 1997.

Pringle, H., 'The Slow Birth of Agriculture' in *Science* vol. 282, 20 Nov. 1998.

Prothero, D. R., *The Eocene-Oligocene Transition*, Columbia University Press, New York, 1994.

Prothero, D. R. & Beggren, W. A. eds, *Eocene-Oligocene Climatic and Biotic Evolution*, Princeton University Press, Princeton, 1992.

Prothero, D. R. & Emry, R. J. eds, *The Terrestrial-Oligocene Transition in North America*, Cambridge University Press, Cambridge, 1996.

Pyne, S. J., *Fire in America: A Cultural History of Wildland and Rural Fire*, University of Washington Press, Seattle, 1982.

Reisner, M., *Cadillac Desert: The American West and Its Disappearing Water*, Viking, New York, 1986.

Roberts, R. G., 'Luminescence Dating in Archaeology: From Origins to Optical', *Radiation Measurements* vol. 27, 1998,

Rose, K. D., 'On the Origin of the Order Artiodactyla', *Proceedings of the National Academy of Science USA* vol. 93, 1996.

Sauer, C. O., *Sixteenth Century North America*, University of California Press, Berkeley, 1975.

Schaal, S. & Ziegler, W. eds, *Messel: An Insight into the History of Life and of the Earth*, trans. M. Shaffer-Fehre, Clarendon Press, Oxford, 1992.

Schlereth, T. J., *Victorian America: Transformations in Everyday Life*, Harper-Perennial, New York, 1991.

Schultz, J. C. & Floyd, T., 'Desert Survivor' in *Natural History* vol. 108 (1), 1999.

Schultz, P. H. & D'Hondt, S. D., 'Cretaceous-Tertiary (Chicxulub) Impact Angle and Its Consequences' in *Geology* vol. 24, 1996.

Shackleton, N. J., 'Paleogene Stable Isotope Events' in *Paleogeography, Paleoclimatology, Paleoecology* vol. 57, 1986.

Sheehan, P. M., Fastovsky, D. E., Hoffmann, R. G., Berghaus, C. B. & Gabriel, D. L., 'Sudden Extinction of the Dinosaurs: Latest Cretaceous, Upper Great Plains, USA' in *Nature* vol. 254, 1991.

Shenkman, R., *Legends, Lies and Cherished Myths of American History*, HarperPerennial, New York, 1989.

Shukolyukov, A. & Lugmair, G. W., 'Isotopic Evidence for the Cretaceous-Tertiary Impactor and Its Type' in *Science* vol. 282, 1998.

Sibley, C. G. & Ahlquist, J. E., *Phylogeny and Classification of Birds*, Yale University Press, New Haven, 1990.

Simpson, G. G., *Splendid Isolation: The Curious History of South American Mammals*, Yale University Press, New Haven, 1980.

Smit, J., 'Extinction and Evolution of Planktonic Foraminifera at the Cretaceous-Tertiary Boundary after a Major Impact', *Geological Society of America special paper no. 190*, 1994.

Smith, J., *The True Travels, Adventures and Observations of Captain John Smith in Europe, Asia, Africa and America, and the General History of Virginia, New England and the Summer Isles*, books 1–3, Cambridge University Press, Cambridge, 1608–1630, republished 1908.

Stanford, D. J. & Day, J. S. eds, *Ice Age Hunters of the Rockies*, Denver Museum of Natural History and University Press, Colorado, 1992.

Steadman, D. W., & Mead, J. I. eds, *Late Quaternary Environments and Deep History: A Tribute to Paul S. Martin*, Mammoth Site Hot Springs, South Dakota, scientific papers, 1995.

Steadman, D. W., Stafford, T. W. & Funk, R. E., 'Nonassociation of Paleoindians with AMS-Dated Late Pleistocene Mammals from the Dutchess Quarry Caves, New York' in *Quaternary Research* vol. 47, 1997.

Stehli, F. G & Webb, S. D. eds, *The Great American Biotic Interchange*, Plenum Press, New York, 1985.

Stephan, S. A., *The Passing of the Passenger Pigeon*, Game Stories no. 18–19, 1932.

Stewart, K. M. & Seymour, K. L. eds, *Palaeoecology and Palaeoenvironments of Late Cenozoic Mammals*, University of Toronto Press, Toronto, 1996.

Steyn, M. & Henneberg, M., 'Pre-Columbian Presence of Treponemal Disease: A Possible Case from Iron Age Southern Africa' in *Current Anthropology* vol. 36, 1995.

Sundell, K. S., 'Oreodonts: Large Burrowing Mammals of the Oligocene', Society of Vertebrate Paleontology Abstracts, 1997.

Tambussi, C. P., 'Fororracoideos, las grandes aves carnívoras de la Patagonia de antaño' in *Museo*, 1999.

Teale, E. W., *Autumn across America*, St Martin's Press, New York, 1956.

Tedford, R., 'History of Cats and Dogs' in *A View from the Fossil Record: In Nutrition and Management of Dogs and Cats*, Ralston Purina, St Louis, 1978.

Tenner, E., 'Differentiating the Big Three' in *Princeton Alumni Weekly*, 7 Nov., 1984.

Tenner, E., 'The Technological Imperative' in *Wilson Quarterly* vol. 19 (1), 1995.

Thewissen, J. G. M., 'Evolution of Paleocene and Eocene Phenacodontidae (Mammalia: Condylarthra)' in *Papers in Paleontology* vol. 29, 1990.

Thornton, I., *Krakatau: The Destruction and Reassembly of an Island Ecosystem,* Harvard University Press, Cambridge, 1996.

Ting, S., 'Paleocene and Early Eocene Land Mammal Ages of Asia' in *Dawn of the Age of Mammals in Asia,* Beard, K. C. & Dawson, M. R. eds, *Bulletin of Carnegie Museum of Natural History* no. 34, Pittsburgh, 1998.

Trollope, F., *Domestic Manners of the Americans,* Alan Sutton, Dover, New Hampshire, 1832.

Turner, F. J., *The Frontier in American History,* first pub. Henry Holt & Co., New York, 1893, reprinted 1947.

2000 IUCN Red List Threatened Species (www.redlist.org).

Voorhies, M. R., 'Ancient Ashfall Creates Pompeii of Ancient Animals' in *National Geographic,* Jan 1981.

Voorhies, M. R. & Perkins, M. E., 'Odocoilene Deer (?*Bretzia*) from the Santee Ash Locality, Late Hemphillian, Nebraska: Oldest New World Cervid?', Society of Vertebrate Paleontology Abstracts, 1998.

Wallace, A. R., *The Geographical Distribution of Animals: With a Study of the Relations of Living and Extinct Faunas as Elucidating Past Changes of the Earth's Surface,* Harper, New York, 1876.

Ward, P. D., *The Call of Distant Mammoths,* Copernicus Springer-Verlag, New York, 1997.

Watts, W. A. & Hansen, B. C. S., 'Pre-Holocene and Holocene Pollen Records of Vegetation History from the Florida Peninsula and Their Climatic Implications', in *Palaeogeography, Palaeoclimatology, Palaeoecology* vol. 109, 1994.

Webb, S. D., 'A History of Savanna Vertebrates in the New World' (part 1) *Annual Review of Ecology and Systematics* vol. 8, 1977.

Webb, S. D., 'The Rise and Fall of the Late Miocene Ungulate Fauna in North America' in *Coevolution,* M. H. Nitecki ed., University of Chicago Press, Chicago, 1983.

Webb, T., 'The Appearance and Disappearance of Major Vegetational Assemblages: Long-Term Vegetational Dynamics in Eastern North America' in *Vegetatio* vol. 69, 1987.

Wilcove, D. S., *The Condor's Shadow: The Loss and Recovery of Wildlife in America,* Freeman & Co., New York, 1999.

Wilson, J., *The Earth Shall Weep: A History of Native America,* Atlantic Monthly Press, New York, 1999.

Wilson, M., 'Archaeological Kill Site Populations and the Holocene Evolution of the

Genus *Bison*' in *Bison Evolution and Utilisation*, L. Davis & M. Wilson eds, *Plains Anthropologist* memoir no. 14, 1978.

Wing, S. L., 'Eocene & Oligocene Floras and Vegetation of the Rocky Mountains' in *Annals of the Missouri Botanical Garden* vol. 74, 1987.

Winslow, E., 'Good Newes from New England: A True Relation of Things Very Remarkable at the Plantation of Plimoth in New England', (facsimile) Applewood Books, Bedford, 1624.

Winter, W. H., 'Elephant Behaviour' in *East African Wildlife Journal* no. 2, 1964.

Winthrop, J., *The Journal of John Winthrop 1630–1649*, abridged edn, R. S. Dunne & L. Yeandle eds, Harvard University Press, Cambridge, 1996.

Withington, A. F., *Towards a More Perfect Union: Virtue and the Formation of American Republics*, Oxford University Press, New York, 1991.

Witmer, L. W. & Rose, K. D., 'Biomechanics of the Jaw Apparatus of the Gigantic Eocene Bird Diatryma: Implications for Diet and Mode of Life' in *Paleobiology* vol. 17, 1991.

Wolfe, J. A., 'Tertiary Climatic Fluctuations and Methods of Analysis on Tertiary Floras' in *Palaeogeography, Palaeoclimatology, Palaeoecology* vol. 9, 1971.

Wolfe, J. A., 'A Paleobotanical Interpretation of Tertiary Climates in the Northern Hemisphere' in *American Scientist* vol. 66, 1978.

Wolfe, J. A., 'Distributions of Major Vegetation Types during the Tertiary' in *The Carbon Cycle and Atmospheric CO2: Natural Variations, Archean to Present*, E. T. Sundquist & W. S. Broeker eds, American Geophysical Union Geophysical Monographs vol. 32, 1985.

Wolfe, J. A., 'Tertiary Floras and Paleoclimates of the Northern Hemisphere' in *Land Plants: Notes for a Short Course*, T. W. Broadhead ed., University of Tennessee Dept. Geological Sciences studies in Geology 15, 1986.

Wolfe, J. A., 'Late Cretaceous–Cenozoic History of Deciduousness and the Terminal Cretaceous Event' in *Paleobiology* vol. 13, 1987.

Wolfe, J. A., 'North American Non-Marine Climates and Vegetation during the Late Cretaceous' in *Palaeogeography, Palaeoclimatology, Palaeoecology* vol. 61, 1987.

Wolfe, J. A. & Upchurch, G. R., 'Leaf Assemblages across the Cretaceous-Tertiary Boundary in the Raton Basin, New Mexico and Colorado', *Proceedings National Academy Science* vol. 84, 1987.

Wynn, J. C. & Shoemaker, E. M., 'The Day the Sands Caught Fire' in *Scientific American*, Nov. 1998.

ACKNOWLEDGMENTS

This book was researched and partly written while I was chair of Australian Studies at Harvard University during the 1998–99 academic year. The Committee on Australian Studies gave me every encouragement, and special thanks are due to Janet Hatch, Harold Bolitho and Bernard Bailyn, all of whom liberally extended hospitality and assistance. Fuzz Crompton and Farish Jenkins made a home for me in the Department of Organismic and Evolutionary Biology and, along with David Haig, Yvonne Parsons and Chris Norris, provided companionship and critical assessment of many of my ideas.

To the staff of the Ernst Mayr Library I owe a greater debt than I can repay, for they cheerfully aided me in ways far above and beyond the call of duty as I pursued ever more obscure works. My Harvard students are owed my thanks; their fresh perspectives and enthusiasm made researching this book and teaching the course from which it grew a pleasure.

Jared Diamond kindly gave his time to read the manuscript even while overburdened by his own obligations. Thys ter Horst's reading of an early draft was a test of patience, and the drafts that followed benefited from his breadth of knowledge and endless common sense.

Richard Tedford of the American Museum of Natural History, Patricia Vickers-Rich of Monash University, Thomas Rich of the Museum of Victoria, Peter Mathews and John Salmond of La Trobe University, Brian McGowran of the University of Adelaide, David Webb and Andy Hemmings of the Florida Museum of Natural History and Russ Graham of the Denver Museum read the manuscript, providing insights and corrections in their areas of expertise. Mark Westoby of Macquarie University and his students Martin Henery, Cassia Read, Peter Coppinger, Peter Vesk, Barbara Rice and Angela Moles critiqued the manuscript as part of their course work. Christine Janis of Brown University, who also read a draft of this work, is owed a special debt of gratitude. As senior editor of *Evolution of Tertiary Mammals of North America, Volume 1*, she has produced a work that assisted me greatly in my research, and will stand as a definitive reference for decades to come.

Writing this book would not have been possible without the enthusiasm and patience of my spouse Alexandra Szalay, who read the manuscript in several manifestations and brought my attention to useful references.

Finally my thanks go to my editors Michael Heyward and Melanie Ostell at Text Publishing, and Morgan Entrekin and Andrew Miller at Grove Atlantic, who together artfully shaped what was sometimes an unruly growth.

INDEX

(Common names are used where they exist. Most extinct species are listed under scientific names.)